ENVIRONMENTAL RESEARCH ADVANCES

CLIMATE, ENERGY, AND ENVIRONMENT

ISSUES, ANALYSES, AND DEVELOPMENTS

VOLUME 1

ENVIRONMENTAL RESEARCH ADVANCES

Additional books in this series can be found on Nova's website under the Series tab.

Additional E-books in this series can be found on Nova's website under the E-book tab.

ENERGY SCIENCE, ENGINEERING AND TECHNOLOGY

Additional books in this series can be found on Nova's website under the Series tab.

Additional E-books in this series can be found on Nova's website under the E-book tab.

ENVIRONMENTAL RESEARCH ADVANCES

CLIMATE, ENERGY, AND ENVIRONMENT

ISSUES, ANALYSES, AND DEVELOPMENTS

VOLUME 1

JAKE S. DIAZ
EDITOR

New York

For permission to use material from this book please contact us:
Telephone 631-231-7269; Fax 631-231-8175
Web Site: http://www.novapublishers.com

Library of Congress Cataloging-in-Publication Data

ISBN: 978-1-62948-407-5

Published by Nova Science Publishers, Inc. † New York

CONTENTS

PREFACE

This first volume of "Climate, Energy, and Environment: Issues, Analyses, and Developments" presents current topical research in the study of the challenges that affect whether actions to preserve forests could make a cost-effective contribution to reducing global emissions of greenhouse gases; the potential of ocean fertilization for climate change mitigation; the energy and economic impacts of U.S. federal energy and water conservation standards adopted from 1987 through 2011; and Klamath Basin settlement agreements.

Chapter 1 – Human activities produce large amounts of greenhouse gases (GHGs), primarily carbon dioxide (CO_2), and thus contribute to global warming. The use of fossil fuels is the primary source of CO_2 emissions, but the removal of trees from forested land has also contributed. Mature forests, having absorbed CO_2 from the atmosphere while growing, store carbon in wood, leaves, and soil. That carbon is released when people clear forested land and destroy the wood. From 2000 to 2005, the loss of forests, primarily in tropical developing countries, accounted for approximately 12 percent of global GHG emissions. Slowing or halting deforestation in developing countries is a potentially low-cost way to help reduce global GHG emissions. For that potential to be realized, however, substantial challenges would need to be addressed—by providing technical and financial assistance to governments, by creating demand from private markets, or both.

Chapter 2 – Page 636 of House Report 111-366 that accompanies the Consolidated Appropriations Act of 2010 (Public Law 111-117) calls for the National Oceanic and Atmospheric Administration (NOAA) to "provide a report on the potential of ocean fertilization for climate change mitigation" to the House and Senate Committees on Appropriation within 60 days of enactment of the Act. Climate change mitigation includes any efforts to reduce climate change including reducing emissions of heat-trapping gases and particles, and increasing removal of heat-trapping gases from the atmosphere.

The oceans contain about 50 times as much carbon dioxide (CO_2) as the atmosphere, comprising around 38,118 billion metric tons of carbon compared to 762 billion metric tons in the atmosphere. What allows the oceans to store so much CO_2 is the fact that when CO_2 dissolves in surface seawater, it reacts with a vast reservoir of carbonate ions to form bicarbonate ions. This reaction effectively removes the dissolved gas form of CO_2 from the surface water, allowing the water to absorb more gas from the overlying air. This process, in combination with large-scale ocean circulation, has resulted in the transfer of between a quarter and a third of human-induced emissions of CO_2 from the atmosphere into the ocean since the beginning of the industrial revolution. Ocean biology enhances the ocean's ability to absorb CO_2 from the atmosphere as follows: plants in the ocean, mostly microscopic floating plants called phytoplankton, absorb CO_2 and nutrients when they grow, packaging them into organic material. Most of this organic material is consumed by other organisms and eventually recycled into CO_2 and nutrients near the surface. However, about 1-2% sinks into the deep ocean where it is converted back into CO_2 and nutrients by the action of bacteria. The resulting CO_2 and associated nutrients can remain out of contact with the atmosphere for hundreds of years, until ocean circulation brings the deep water to the surface and the cycle can begin again. Recently, three U.S.-based companies have proposed fertilizing the ocean with iron and other essential nutrients to enhance the tendency of ocean biology to transfer CO_2 from the surface ocean to the deep ocean. These companies seek to deploy ocean fertilization as a climate change mitigation measure. In order for ocean fertilization to lead to climate change mitigation, three criteria must be met: (a) ocean fertilization must lead to increased growth of phytoplankton, packaging carbon and nutrients together into organic material; (b) this organic material must be transferred into the deep ocean so that it does not simply get recycled near the surface releasing its carbon back to the atmosphere; and (c) this transfer of carbon from the surface ocean to the deep ocean must result in a compensating transfer of carbon from the atmosphere into the surface ocean.

This report reviews two decades of ocean fertilization research that has focused primarily on adding the micronutrient iron to regions of the ocean that are iron-poor. A limited amount of research has focused on adding the nutrients nitrogen or phosphorus to areas that primarily lack those nutrients. However, since very small amounts of iron are required by phytoplankton to make organic matter (1 atom for every 50,000 to 200,000 atoms of carbon) compared to the nutrients nitrogen (1 atom for every 16 atoms of carbon) or phosphorus (1 atom for every 106 atoms of carbon), iron fertilization has been seen as the most feasible approach for altering the ocean's carbon cycle on a significant scale and has been the focus of scientific and commercial interest in the United States. This report focuses on the potential for fertilizing the ocean with iron to lead to climate change mitigation. Ocean fertilization research has made an extremely valuable contribution to the scientific understanding of the ocean carbon cycle and its role in the global carbon cycle on time scales ranging from glacial episodes thousands of years in the Earth's past to today's changing climate. All reported ocean fertilization experiments have resulted in increased growth of phytoplankton (criterion (a)). However, the resulting transfer of organic material from the surface ocean into the deep ocean (criterion (b)) and the compensating transfer of carbon into the ocean from the atmosphere (criterion (c)) have not been verified. Thus, the research undertaken to date has not demonstrated that ocean fertilization is a mature technology for climate change mitigation. Modeling studies also suggest that the maximum possible impact of ocean fertilization on atmospheric CO2 concentrations is likely to be a small fraction of what is needed to stabilize CO2 concentrations at twice the preindustrial concentration, a general target for avoiding catastrophic climate change. There are also environmental risks associated with ocean fertilization. Adding nutrients to the ocean causes changes in the structure and functioning of marine ecosystems, and these changes may result in unintended negative consequences. Scientists hypothesize that such consequences may include: decreases in productivity in ocean regions remote from the fertilization site; alteration of the relative abundance, size structures and diversity of higher trophic levels, including (but not limited to) economically important species; increases in deep ocean hypoxia or anoxia; increases in the oceanic production of the greenhouse gases nitrous oxide and methane; generation of toxin-producing algae; introduction of toxic chemicals at the fertilization site (depending on the composition of the fertilization material); and increases in ocean acidification with associated increased impacts in sub-surface ecosystems already vulnerable to ocean acidification. Concern about the environmental consequences of proposed large-scale ocean fertilization experiments has triggered the development of an international regulatory mechanism through the 1972 Convention on the Prevention of Marine Pollution by Dumping of Wastes and Other Matter, to which the United States is a Party. NOAA executed an agreement with the National Academy of Sciences to establish a committee on America's Climate Choices, as required under the Department of Commerce Appropriations Act of 2008 (Public Law 110-161). This committee established four expert panels, including a Panel on Limiting the Magnitude of Future Climate Change. The Panel has described, analyzed, and assessed strategies for reducing the net future human influence on climate, including both technology and policy options. The Panel's report, released on May 19 2010, provided Congress with information on a wide range of climate mitigation options. The present report assesses the climate mitigation potential of a single proposed technology, ocean fertilization, as requested by Public Law 111-117.

Chapter 3 – The Lacey Act was enacted in 1900 to prevent hunters from illegally killing game in one state and escaping prosecution by crossing state lines. It has evolved into a law that prohibits import, export, transport, purchase, or sale of species when that action would violate state, federal, tribal, or foreign law. Congress amended the Lacey Act most recently in 2008, expanding the reach of the act to include timber and timber products. Implementation of the 2008 Amendments has proved controversial, and the Department of Agriculture Animal and Plant Health Inspection Service (APHIS) initially delayed implementing the act's new declaration requirements for importing wood products.

Some find the Lacey Act puzzling. While people charged with violating the act are charged with violating a U.S. law, that prosecution is premised on a violation of another law, sometimes the law of another country. That has led some to claim that the United States is enforcing the laws of another country. U.S. conservation laws (such as the Lacey Act), however, have long protected species and habitats even outside of the United States. Worldwide conservation was one reason for expanding Lacey Act coverage to more plants in 2008. Preserving U.S. timber jobs and prices was another reason. However, the 2008 Amendments to the Lacey Act allow enforcement of foreign laws that are not directly related to conservation or U.S. jobs, such as failure to pay foreign stumpage fees, or shipping wood in violation of a country's export restrictions. After search warrants were executed by the Department of the Interior Fish and Wildlife Service (FWS) against Gibson Guitar Corp. of Nashville, TN, in August 2011, based on the possible illegal import of wood from India, the 112th Congress took another look at whether the 2008 Amendments achieve the goals of the Lacey Act.

In July 2012, Gibson settled charges that imports in 2008 and 2009 had violated the Lacey Act, agreeing to pay a $300,000 penalty, make a $50,000 donation to a wildlife nonprofit group, and withdraw its claims to recover wood forfeited during the Madagascar investigation. The United States returned the wood from India seized in 2011. As part of the settlement, the

United States agreed that Indian trade law, on which the alleged 2011 violations were based, was unclear. However, Gibson admitted that it had continued to import ebony from Madagascar after learning that it was illegal.

Chapter 4 – This paper presents estimates of the key impacts of Federal energy and water conservation standards adopted from 1987 through 2011. The standards for consumer products and commercial and industrial equipment include those set by legislation as well as standards adopted by DOE through rulemaking. In 2011, the standards saved an estimated 3.2 quads of primary energy, which is equivalent to 3% of total U.S. energy consumption. The savings in operating costs for households and businesses totaled $42 billion. The average household saved $309 in operating costs as a result of residential and plumbing product standards. The estimated reduction in CO_2 emissions associated with the standards in 2011 was 176 million metric tons, which is equivalent to 3 percent of total U.S. CO_2 emissions. The estimated cumulative energy savings over the period 1990-2070 amount to 176 quads. Accounting for the increased upfront costs of more-efficient products and the operating cost (energy and water) savings over the products' lifetime, the standards have a projected cumulative net present value (NPV) of consumer benefit of between $938 billion and $1,279 billion, using 7 percent and 3 percent discount rates, respectively. The water conservation standards, together with energy conservation standards that also save water, reduced water use by 1.5 trillion gallons in 2010, and will achieve cumulative water savings by 2040 of 51.4 trillion gallons. The estimated consumer savings in 2010 from reduced water use amounted to $10.8 billon. Standards set by legislation account for 47 percent of total cumulative primary energy savings and 55 to 66 percent of the total cumulative NPV (depending on which discount rate is used). Standards set by DOE rulemakings account for 53 percent of total cumulative primary energy savings and 34 to 45 percent of the total cumulative NPV.

Chapter 5 – The Klamath River Basin on the California-Oregon border is a focal point for local and national discussions on water allocation and species protection. Previously, water and species management issues have exacerbated competition and generated conflict among several interests—farmers, Indian tribes, commercial and sport fishermen, federal wildlife refuge managers, environmental groups, and state, local, and tribal governments. Drought conditions and a recent call for water by senior water rights holders have again brought these issues to the forefront.

In 2010, the Secretary of the Interior and the governors of Oregon and California, along with multiple interest groups, announced the result of multi-year negotiations in an effort to resolve long-standing issues in the basin: two interrelated agreements, supported by the federal government and signed by the two states and numerous other parties. These agreements, known as the Klamath Basin Restoration Agreement (KBRA) and the Klamath Hydroelectric Settlement Agreement (KHSA), together aim to provide for water deliveries to irrigators and wildlife refuges, fish habitat restoration, and numerous other related actions. Generally, the KBRA provides for actions intended to restore Klamath fisheries and for assurances for water deliveries to wildlife refuges and federal project irrigators under certain circumstances, among many other things. The KHSA lays out a process that could potentially lead to the removal of four privately owned dams on the Klamath River. This dam removal would be one of the largest and most complex projects of its kind ever undertaken.

Some parts of the Klamath Agreements are already being carried out under existing authorities. Studies to inform a Secretarial Determination on dam removal under the KHSA have been completed, and some restoration actions have already begun. However, new authorization from Congress is required in order for the most significant components of the agreements to be implemented. Other ongoing processes in the Klamath Basin, such as the recent water rights adjudication by the State of Oregon, may result in significant changes in water allocations in the basin with or without federal legislation to authorize the Klamath Agreements. However, these developments are related.

The Klamath Agreements were originally set to expire in 2012 if no authorizing legislation was enacted, but have been extended to 2014. Legislation in the 11^{2th} Congress would have authorized the agreements, including hundreds of millions of dollars in federal actions under the KBRA and authorization for the Secretary of the Interior to make a "final" dam removal recommendation under the KHSA. No such legislation has been proposed in the 113th Congress.

Congressional consideration of the Klamath Agreements could include whether the federal government should act beyond its current activities in the Klamath Basin, the type of strategies proposed in the agreements and their cost, and the potential for the approach in the Klamath Basin to serve as precedent for the resolution of similar conflicts in other parts of the country.

In: Climate, Energy, and Environment. Vol. 1 ISBN: 978-1-62948-407-5
Editor: Jake S. Diaz

Chapter 1

DEFORESTATION AND GREENHOUSE GASES*

Natalie Tawil

PREFACE

Emissions of carbon dioxide, a greenhouse gas, come primarily from the consumption of fossil fuels and from deforestation. The destruction and degradation of forestland, caused mainly by expanded agricultural activity in tropical developing countries, currently accounts for an estimated 12 percent of global greenhouse gas emissions.

This Congressional Budget Office (CBO) study, prepared at the request of the Chairman of the Senate Committee on Foreign Relations, examines challenges that affect whether actions to preserve forests could make a cost-effective contribution to reducing global emissions of greenhouse gases. It also discusses the pros and cons of approaches that the United States and other developed countries could take to promote forest preservation. In keeping with CBO's mandate to provide objective, impartial analysis, the study contains no recommendations.

Douglas W. Elmendorf
Director

---

* This is an edited, reformatted and augmented version of The Congressional Budget Office, Publication No. 4216, dated January 2012.

SUMMARY

Human activities produce large amounts of greenhouse gases (GHGs), primarily carbon dioxide (CO_2), and thus contribute to global warming. The use of fossil fuels is the primary source of CO_2 emissions, but the removal of trees from forested land has also contributed.

Mature forests, having absorbed CO_2 from the atmosphere while growing, store carbon in wood, leaves, and soil. That carbon is released when people clear forested land and destroy the wood. From 2000 to 2005, the loss of forests, primarily in tropical developing countries, accounted for approximately 12 percent of global GHG emissions.

Slowing or halting deforestation in developing countries is a potentially low-cost way to help reduce global GHG emissions. For that potential to be realized, however, substantial challenges would need to be addressed—by providing technical and financial assistance to governments, by creating demand from private markets, or both.

Challenges in Reducing Forest-Based Emissions

If actions to support forest preservation are to play a cost-effective role in a significant international effort to reduce global GHG emissions, three broad challenges would have to be met:

- Obtaining useful measurements of changes in the amount of carbon stored in forests,
- Structuring incentives to reduce total forest-based emissions, and
- Improving governance in developing countries.

Measuring Changes in Carbon Storage

Establishing programs to reduce emissions of greenhouse gases and assessing the effectiveness of those programs require methods for measuring emissions. In some cases, measuring them is easy—the electric power industry, for example, can use systems that directly, continuously, and accurately monitor CO_2 emissions.

Measuring emissions resulting from deforestation is more complicated, however, because such emissions depend on the amount of deforestation and the carbon content of the wood that has been destroyed. Researchers can combine remote-sensing data about the amount of deforestation with information about the carbon content of the wood gleaned from on-the-ground inventories of the number and size of trees in sample areas to make such measurements. Most developing countries would need to improve their technical capabilities to process remote-sensing data and conduct inventories in order to effectively implement any program aimed at reducing carbon emissions from deforestation.

Structuring Incentives to Reduce Total Forest-Based Emissions

To reduce total forest-based emissions worldwide, the design of preservation programs must consider not only how much additional preservation would result but also how much "leakage" would occur—that is, how much of a forest program's direct reductions in GHG emissions would be negated by additional GHG releases elsewhere.[1]

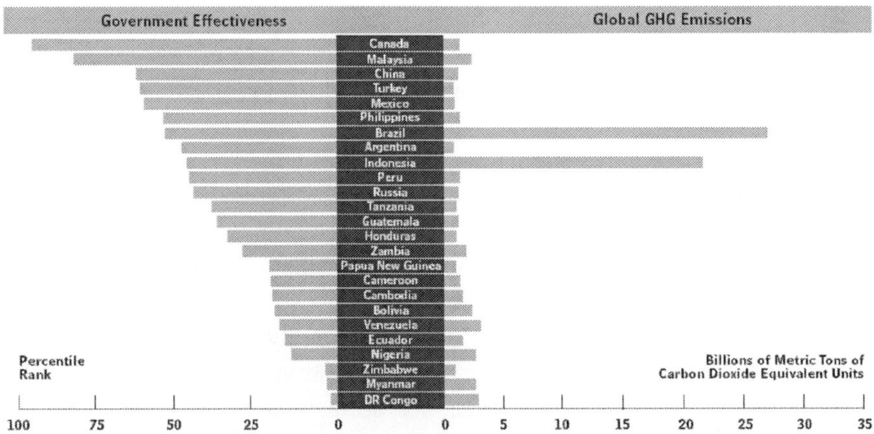

Sources: Congressional Budget Office based on data from World Bank, "Worldwide Governance Indicators" (2009), http://info .worldbank.org/governance/wgi/sc_ country.asp; and World Resources Institute, "Climate Analysis Indicators Tool (CAIT)" (2011), http://cait.wri.org.

Notes: "Government effectiveness" is a governance indicator reported by the World Bank. It measures the quality of public services, the quality of the civil service and the degree of its independence from political pressures, the quality of policy formulation and implementation, and the credibility of the government's commitments to such policies.

GHG = greenhouse gas; DR Congo = Democratic Republic of the Congo.

Summary Figure 1. Government Effectiveness in Countries Responsible for 95 Percent of Global Forest-Based Emissions, 1990 to 2005.

For example, a program that compensates people for preservation in one location might prompt a decline in the clearing of forested land for agriculture or timber production in that area, thus reducing supplies of those commodities and raising their prices. Higher prices, in turn, could encourage uncompensated landowners elsewhere to clear forests to produce agricultural products or timber to sell at the higher prices. Consequently, programs might need to compensate not only new preservation efforts aimed at threatened forests but also the continued preservation of forests that would not be threatened in the program's absence. Leakage reduces a program's cost-effectiveness, and significant leakage might negate the cost advantages that using forest preservation has in comparison with other approaches to achieving GHG reductions.

Improving Governance in Developing Countries

Weak governance—the inability to successfully design and implement policies to achieve stated objectives— undermines any efforts to use forest preservation programs to produce verifiable reductions in greenhouse gases. Even if they are motivated to participate in such programs, agencies in some developing countries may have inadequate authority for that task and may lack effective mechanisms for negotiating and distributing compensation to those who preserve forests. Also, the rights to any potential benefits from preserving forest resources may be poorly defined, making the gains from deforestation for agricultural and timber production and the use of wood for fuel more certain than the gains from preserving forests. Finally, government corruption and political instability can undermine laws that promote preservation.

Improving governance may be the most intractable of the three challenges. Of the 25 countries with the largest forest-based emissions in recent years, which together produced 95 percent of such emissions globally, nearly three-quarters rank in the bottom half of all countries on a key indicator of a country's ability to govern—government effectiveness. That indicator measures, for example, the quality of policy formulation and implementation and the credibility of the government's commitments to its policies. The World Bank, which tracks and reports measures of governance, rates the two largest emitters of forest-based CO_2—Brazil and Indonesia—at roughly the 50th percentile in terms of government effectiveness (see Summary Figure 1).

Policy Approaches for Reducing Forest-Based Emissions

Approaches the United States and other developed countries could take to encourage forest preservation in developing countries fall into two broad categories:

- Providing financial and technical assistance to governments interested in preserving forests and
- Creating demand in private markets for reductions in forest-based greenhouse gas emissions.

The two types of policies might work best together. The viability of markets, for example, may depend on having in place a reliable program for achieving measurable reductions in forest-based emissions—the type of program that financial and technical assistance can help establish.

Assistance to Governments

Financial and technical assistance can help overcome some of the challenges of pursuing forest preservation. It can help support advances in measuring and monitoring changes in forest carbon, help ensure that developing countries have access to the technologies for doing so, and also help counter leakage by offering incentives for achieving global reductions in forest-based emissions. Given uncertain funding and the challenges of improving governance in developing countries, the United States and other developed countries could consider focusing efforts on selected countries—for example, Brazil and Indonesia—that have relatively reliable governance, that are rich in remaining forest resources, and whose experiences could inform subsequent policy development.

Markets for Forest-Based Emissions

The United States and other developed countries could also generate resources for reducing forest-based GHG emissions by creating demand in private markets for such reductions. They could do that by establishing cap-and-trade programs or by taxing GHG emissions and providing tax credits for those who fund forest preservation activities. The potential for forest preservation in developing countries to lower the private-sector costs of achieving a goal for global GHG reductions might motivate substantial funding from private sources.

Deforestation and Greenhouse Gases

Emissions of greenhouse gases (GHGs) caused by human activity contribute to climate change. In particular, experts attribute most of the warming of the climate to emissions of carbon dioxide (CO_2). Although the use of fossil fuels for energy is the primary source of CO_2 emissions, the loss of forests is also a major contributor.

Forests affect the amount of CO_2 in the atmosphere in a number of ways.[1] As forests grow, they remove CO_2 from the atmosphere and absorb carbon into wood, leaves, and soil, where it can be stored for an extended period. When forests are cleared, stored carbon may be released into the atmosphere, depending in part on how much of the wood is destroyed. For example, using fires to clear forested land for agricultural production or other uses produces more emissions than does felling timber for wood products, which if disposed of in landfills at the end of their use will continue to store carbon.

The vast amount of carbon stored in forests worldwide indicates the important role of forests in climate change. The trees in forests are estimated to store the equivalent of roughly 760 billion metric tons of CO_2 worldwide—over one hundred times the United States' emissions of CO_2 and other greenhouse gases in 2009.[2] Most of the locations that are rich in forest carbon are in tropical countries (see the dark green areas in Figure 1). Preserving forests—by slowing or eliminating the loss of forests and the selective removal of trees (a process referred to as degradation)—would reduce CO_2 emissions.

Carbon storage in forests also can be increased by growing trees on previously unforested land (the brown areas shown in Figure 1). Globally, however, planting trees would contribute less to reducing concentrations of greenhouse gases in the atmosphere than would forest preservation. The potential for new forests to store carbon is limited by biological factors (such as the soil and climate conditions on available land) and economic factors (such as the benefits associated with current uses of that land that would have to be forgone if forests were planted).

For the same reasons, other forest-based activities—such as restoring degraded forests and practicing sustainable forest management—have even less potential to reduce concentrations of greenhouse gases in the atmosphere. Restoring degraded forests involves techniques that stimulate the natural regeneration of trees by reducing the intensity of grazing or planting selected areas with mixed tree species to help promote growth. Sustainable forest management, implemented on lands dedicated to timber production, involves practices such as reduced-impact logging, which includes limiting the slope of roads to prevent erosion, careful planning to minimize road building, and allowing trees to grow larger before harvesting them.

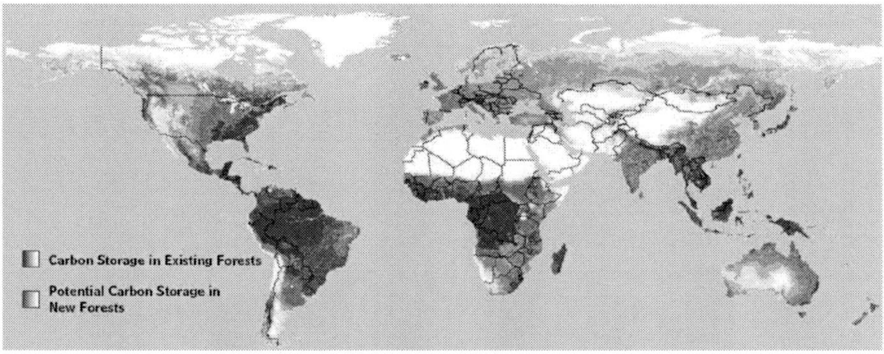

Source: Congressional Budget Office adapted from Adrian Deveny, Janet Nackoney, Nigel Purvis, and others, *Forest Carbon Index: The Geography of Forests in Climate Solutions* (Washington, D.C.: Resources for the Future and Climate Advisers, December 2009).

Note: The amount of carbon storage is greater in the dark green areas than in the lighter green areas; the potential for carbon storage is greater in the dark brown areas than in the lighter brown areas. Although forest growth is biologically possible in the brown shaded areas, planting trees there would not be a cost-effective way to store carbon on much of the land, given the land's limited potential to support forests and its value for other economic activities.

Figure 1. Estimated Distribution of Carbon Storage in Forests.

This Congressional Budget Office study examines the role of forest preservation in reducing global GHG emissions. It explores the challenges involved in establishing cost-effective programs to achieve such reductions and discusses policy approaches that the United States and other developed countries could take to encourage that effort.

IMPLICATIONS OF DEFORESTATION FOR CLIMATE CHANGE

Human activities lead to emissions of four principal greenhouse gases: carbon dioxide, methane, nitrous oxide, and a group of gases called halocarbons. Concentrations of those gases, which accumulate in the atmosphere, have increased in the industrial era. Carbon dioxide, the most prevalent of the four, accounts for roughly three-quarters of the impact of greenhouse gases on climate, largely because it is a by-product of burning fossil fuels but also because it is released when forests are cleared and the wood is destroyed.

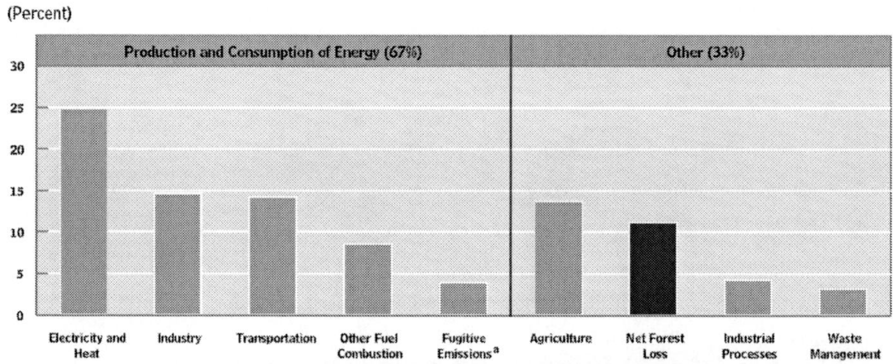

Source: Congressional Budget Office based on data from World Resources Institute, *World Greenhouse Gas Emissions: 2005* (Washington, D.C.: WRI, December 2005), www.wri.org/chart/world-greenhouse-gas-emissions-2005.

Note: The impact on climate of one ton of emissions of a greenhouse gas depends on how that particular gas affects the balance within the Earth's atmosphere between incoming solar radiation and outgoing infrared radiation. To assess the impact of all greenhouse gases on an equal basis, researchers report emissions in terms of carbon dioxide equivalent, or CO_2e. This figure shows the share of emissions from various sources measured in terms of metric tons of CO_2e. A metric ton of CO_2e is the amount of a given greenhouse gas that has the same impact on climate as a metric ton of CO_2. (A metric ton equals 1.1 tons.)

[a] Fugitive emissions are those associated with the production, processing, storage, transmission, and distribution of fossil fuels such as oil, natural gas, and coal (including abandoned underground coal mines).

Figure 2. Estimated Shares of Global Emissions of Greenhouse Gases Caused by Human Activity, by Source, 2005.

One widely cited estimate of the impact of net forest loss during the 1990s, for example, put its share of global GHG emissions—measured in terms of the impact on warming—at 20 percent, within a likely range of 6 percent to 34 percent.[3] That estimate averaged two assessments—one based on ground-level forest inventories (the counting and measuring of trees in sample plots) undertaken by countries and reported to the United Nations' Food and Agriculture Organization (FAO) and the other based on data gathered by remote-sensing instruments. That 20 percent estimate was later revised downward—to 16 percent—reflecting corrected FAO data indicating that the amount of tropical deforestation during that period was less than originally thought.[4]

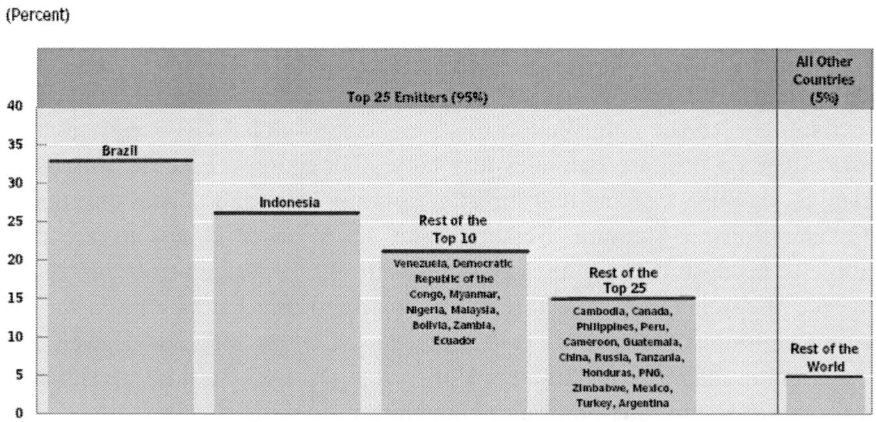

Sources: Congressional Budget Office based on data from World Resources Institute, "Climate Analysis Indicators Tool (CAIT)" (2011), http://cait.wri.org. For details, see World Resources Institute, *CAIT: GHG Sources and Methods* (Washington, D.C.: WRI, November 2010), http://cait.wri.org/downloads/cait_ghgs.pdf; and World Bank, *World Development Report: Development and Climate Change*, Selected Indicators, Table A2 (Washington, D.C.: World Bank, 2010), http://siteresources.worldbank.org/INTWDR2010/Resources/5287678-1226014527953/WDR10-Full-Text.pdf.

Notes: Researchers use regional and global estimates of changes in land use to derive estimates for forest-based emissions of the 25 largest national contributors. They typically base the country estimates on a single data point for each year in the period, and errors associated with the estimates may be substantial.

Emissions of greenhouse gases are measured in terms of carbon dioxide equivalent (CO_2e). A metric ton of CO_2e is the amount of a given greenhouse gas that makes the same contribution to global warming as a metric ton of CO_2. (A metric ton equals 1.1 tons.)

PNG = Papua New Guinea.

Figure 3. Ranking of Countries by Their Estimated Contribution to Global Forest-Based Emissions, 1990 to 2005.

A subsequent assessment, for the 2000–2005 period, produced an even lower estimate of the impact of net forest loss, putting its share at 12 percent of global GHG emissions measured in terms of the impact on warming, within a likely range of 6 percent to 17 percent.[5] The estimate was lower than the estimate in the 1990s mainly because of higher emissions from the combustion of fossil fuels rather than a decline in deforestation.

That 12 percent estimate is similar to the warming impact of global GHG emissions from agriculture, including methane (from animal waste) and nitrous oxide (from the use of fertilizer). That estimate is also similar to estimates of effects from global energy use in industry and transportation (see Figure 2).

Estimates of the contribution of forest loss to global GHG emissions are more uncertain than are estimates for many other sources. Electric utilities, for example, can use systems that directly, continuously, and accurately monitor CO_2 emissions. By contrast, estimates of forest-based emissions are made indirectly because there is no way to measure the emissions from forest loss. Instead, researchers measure forest loss itself and use that as a basis for estimating emissions. Moreover, measurements of forest loss are derived from samples rather than from evaluations of the entire forest, so they are subject to error. For example, one particular challenge in measuring forest-based emissions is effectively accounting for the degradation of forests. Selective removal of certain tree species from mature forests reduces carbon storage but is difficult to detect compared with the changes in land cover associated with more complete clearing of forested land.

CURRENT LOCATIONS AND CAUSES OF DEFORESTATION

About 95 percent of forest-based emissions come from 25 countries, most of which are developing countries in the tropics (see Figure 3). Researchers estimate that about three-quarters of all tropical deforestation stems from the clearing of land for agricultural production, ranging from large-scale businesses to subsistence farming.[6] In South America, for example, large agricultural businesses clear forests to raise beef cattle and to grow soybeans and other crops for domestic and international markets. In Southeast Asia, land is cleared for large-scale farming to produce palm oil and coffee for international markets and for smaller producers to grow crops for domestic markets. Southeast Asia is also an important supplier of timber to international markets.

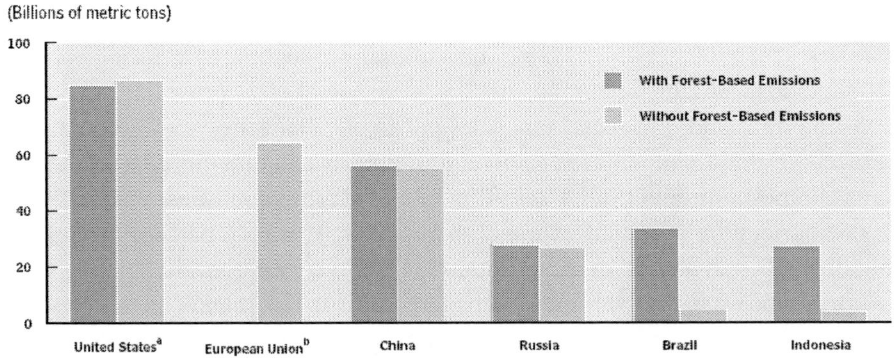

Source: Congressional Budget Office based on data from World Resources Institute, "Climate Analysis Indicators Tool (CAIT)" (2011), http://cait.wri.org.

Notes: A metric ton equals 1.1 tons. CO_2 = carbon dioxide.

[a] In the United States, a net gain in forest area lowered total estimated CO_2 emissions over the 1990–2005 period.

[b] Data on forest-based CO_2 emissions are not available for members of the European Union, but those countries experienced a net gain in forest area over the 1990–2005 period. Data on CO_2 emissions other than forest-based emissions are for the 27 countries that belonged to the European Union in 2005, even if they had not been members for the entire 1990–2005 period. Those countries are Austria, Belgium, Bulgaria, Cyprus, the Czech Republic, Denmark, Estonia, Finland, France, Germany, Greece, Hungary, Ireland, Italy, Latvia, Lithuania, Luxembourg, Malta, the Netherlands, Poland, Portugal, Romania, Slovakia, Slovenia, Spain, Sweden, and the United Kingdom.

Figure 4. Estimated CO_2 Emissions for the Top CO_2-Emitting Nations, 1990 to 2005.

By contrast, most farming in Africa is done on a smaller scale, and it is local communities rather than large businesses that typically encroach on forests as a source of farmland. About 70 percent of Africans depend on small-scale agriculture and the harvesting of natural resources for their livelihood.[7] In addition, 75 percent of urban households and over 90 percent of rural households in sub-Saharan Africa depend on wood, crop residues, and charcoal as their primary source of energy.[8]

In South America and Southeast Asia, deforestation in Brazil and Indonesia is the principal source of forest-based emissions. From 1990 to 2005, forest-based emissions from those two countries were significant enough to place them in the ranks of the top CO_2-emitting nations in the world—along with the United States, China, Russia, and the nations of the European Union combined (see Figure 4). In contrast, on the basis of CO_2 emissions related to fossil fuel alone, Brazil and Indonesia ranked 19th and 21st, respectively.[9]

The economic motivation to clear forests is evident. For individual farmers or landowners, increasing agricultural production by clearing forested land yields tangible benefits in the near term by providing food for consumption and products for sale in local, domestic, and international markets. From a national perspective, agriculture accounts for a large share of gross domestic product (GDP) in tropical developing countries. From 1990 to 2005, agriculture's annual average share of GDP was 7 percent (about $47 billion) in Brazil and 17 percent (about $31 billion) in Indonesia.[10]

In some cases, government subsidies add to the market incentives that encourage agricultural production. Such subsidies reduce the private costs of growing certain kinds of crops, increasing the profits available from that production and thereby providing additional incentives to clear forests. In Indonesia, for example, government policies have supported production of biofuel feedstocks (the main raw materials used in producing biofuels), such as oil palm (the trees from whose fruits palm oil is extracted), by charging below-market interest rates to subsidize loans for developing plantations; by establishing mandatory shares of biofuel use in transportation, industry, and electric power generation; and by funding research and development. The land most suitable for planting oil palm is lowland evergreen tropical rainforest, and about 6 percent of land cleared in Indonesia from 1990 to 2005 was deforested for that purpose.[11] More generally, policies for reducing GHG emissions that promote the expansion of biofuel production but do not assign a value to forest preservation could have the unintended effect of increasing GHG emissions by encouraging deforestation so that land could be cleared for producing biofuel feedstocks (see Box 1).

In contrast, the market incentives for preserving forests are weak. Unlike the benefits of agricultural production, which accrue primarily to producers in the near term, the benefits of preserving forests accrue globally over the longer term by helping to counter climate change and minimize the damage associated with it. Unlike agriculture, efforts to preserve forests lack well-established mechanisms for compensating those who opt for preserving the forest rather than clearing the land. From 2007 to 2010, for example, market-based compensation for climate-related forest preservation averaged an estimated $28 million annually, which was generated through voluntary programs for reducing GHG emissions. That sum is equal to only 0.04 percent of the $78 billion that agriculture contributes annually to the combined GDP of Brazil and Indonesia.[12] And compared with agriculture's annual contribution of approximately $167 billion to the combined GDP of the 21 tropical developing countries experiencing deforestation, the $28 million in compensation is even less significant—just 0.02 percent.

Governments of developing countries and landowners in those countries incur costs when they choose to preserve forests. Governments, for example, incur the costs of establishing programs to monitor forests and enforcing laws that protect them. They may also be reluctant to increase forest preservation because of concerns about whether the distribution of associated benefits among individuals, groups, or regions might create political tensions. Landowners incur the costs of not using the land in other ways—giving up income from the sale of agricultural products, for example. But governments and landowners could also benefit from compensation offered for maintaining the carbon stored in forests and by contributing to efforts to reduce the damage caused by climate change. The balance of costs and benefits would be a significant determinant in their decisions about land use and about participating in an emission-reduction program related to deforestation.

Forests and Cost-Effective Reductions in Greenhouse Gases

Reducing forest-based emissions in developing countries is widely considered to be a relatively inexpensive way to reduce emissions of greenhouse gases even though estimates of the cost-effectiveness of that approach are very uncertain.

Because climate change occurs on a global scale, the climate *benefits* of a reduction in GHG emissions do not depend on either the location of the reduction or the type of source from which it is generated. The *cost* of that reduction, however, depends on both factors. If there were a significant international effort to counter climate change, identifying relatively inexpensive options for reducing global GHG emissions could help lower the total cost of that undertaking.[13]

Box 1. Biofuels and Greenhouse Gas Emissions

Proponents of policies that promote the use of biofuels argue that substituting those renewable energy sources for fossil fuels can help reduce greenhouse gas (GHG) emissions.

Growing biofuel feedstocks—the raw materials, such as corn, sugarcane, switchgrass, oil palms, and soybeans, used to produce biofuel—draws carbon dioxide (CO_2) out of the atmosphere in amounts equal to the emissions produced when the biofuel is burned, whereas using fossil fuels produces net emissions of CO_2. If policies promote the expansion of biofuels but do not place a value on forest preservation, deforestation could spread—increasing associated GHG emissions—as land is cleared for producing the more profitable biofuel feedstocks.

Thus, a key factor in assessing the impact of biofuels on GHG emissions is accounting for the previous use of the land on which the feedstock is grown. If the land was converted from peat, grasses, or forest, that change in land use would boost GHG emissions. Some researchers maintain that it could take decades, or even centuries, before the emission reductions associated with the use of biofuels compensate for the increased emissions from changes in land use. Others conclude that the use of biofuels could result in net reductions in GHG emissions over shorter periods (see the table below).

Estimated Time Required for the Use of Biofuels to Lower Greenhouse Gases When Forestland Is Cleared to Grow Feedstocks

Land Cleared	Location	Years Until Net GHG Reductiona	Study
		Sugarcane Ethanol	
Rainforest	Brazil	45	Searchinger and others
Forest	Brazil	44	Lapola and others
Forest	Brazil	15 to 39	Renewable Fuels Agency
Forest	Brazil	17	Fargione and others
		Palm Biodiesel	
Forest	Brazil	86	Fargione and others
Forest	Malaysia	18 to 38	Renewable Fuels Agency
		Soybean Diesel	
Forest	Brazil	319	Fargione and others
Forest	Brazil	246	Lapola and others

Source: Congressional Budget Office, Using Biofuel Tax Credits to Achieve Energy and Environmental Policy Goals (July 2010), using information from Timothy Searchinger and others, "Use of U.S. Croplands for Biofuels Increases Greenhouse Gases Through Emissions from Land-Use Change," Science, vol. 319 (2008), pp. 1238–1240; David M. Lapola and others, "Indirect Land-Use Changes Can Overcome Carbon Savings from Biofuels in Brazil," Proceedings of the National Academy of Sciences, vol. 107, no. 8

(February 8, 2010), pp. 3388–3393; Renewable Fuels Agency, The Gallagher Review of the Indirect Effects of Biofuels Production (study commissioned by the Secretary of State for Transport, United Kingdom, July 2008); and Joseph Fargione and others, "Land Clearing and the Carbon Debt," Science, vol. 319 (2008), pp. 1235–1238.

Note: GHG = greenhouse gas.

a. The number of years a biofuel would have to be used before the cumulative GHG emissions associated with that biofuel would be lower than the cumulative GHG emissions associated with the use of fossil fuels instead.

The climate policies of other countries and their evolution over time would, however, be critically important to achieving an effective reduction in global emissions of greenhouse gases. Unless a significant percentage of the world's economy restricted emissions, the beneficial effects on climate of any forest-based reduction in emissions could be undermined by emissions from other sources. Although forest preservation can play an important role in many activities—maintaining biodiversity, for example—it is only one strategy among many that would have to be used in a comprehensive global effort to reduce GHG emissions and avert some of the potential damage to the environment and the attendant economic losses associated with climate change.

Cost-Effectiveness of Reducing Forest-Based Emissions

Analyses of cap-and-trade programs that the Congress has considered indicate that reducing forest-based GHG emissions may be a relatively inexpensive way to reduce overall emissions. In such programs, the government sets a cap, or limit, on the amount of GHG emissions that can be produced over a given period. It then distributes rights to emit the gases, or "allowances," by either selling the allowances or giving them away to businesses that must comply with the limits. Businesses can comply with the limits on GHG emissions by submitting one allowance for each ton of CO_2e they emit. Businesses that reduce their emissions at a cost below the price of allowances can sell their rights to emit greenhouse gases, and those with higher costs can buy allowances.

Businesses can also comply with the limits by using "offsets," which substitute alternative, less expensive reductions in emissions for those from the businesses that have to comply with the program. Those alternative reductions can be forest-based (for example, from preserving forests to minimize releases of CO_2), agriculture-based (from reducing emissions of methane from animal waste), or industry-based (from controlling methane emissions from landfills). By changing the mix of activities undertaken to achieve the GHG limits, offsets increase the available supply of allowances, lower their price, and reduce the total cost of meeting the program's specified GHG limits.

The Environmental Protection Agency's (EPA's) estimates of allowance prices under draft cap-and-trade legislation illustrate the potential of offsets, including those from forest preservation, to lower the cost of reducing greenhouse gas emissions. In making its estimates, EPA assumed that actions to reduce such emissions would be widespread and would be taken by developed and developing countries. Allowing offsets from forest preservation would reduce the price of allowances (and the cost of complying with the proposed caps) by 20 percent, EPA estimated.[14] Without such offsets, businesses that had to comply would increase their own emission reductions and purchase more offsets from domestic sources—both of which are more expensive methods of compliance than is forest preservation in developing countries.

Uncertainty About the Cost-Effectiveness of Reducing Forest-Based Emissions

Estimates by EPA and others of the potential cost-effectiveness of a policy that promotes forest-based mitigation are uncertain because quantifying the potential of forests to store carbon involves many factors that are difficult to assess. Those factors include an estimate of what forest-based emissions would be in the absence of the policy, the value of the land in alternative uses over time, the direct costs of undertaking the conservation practices and verifying the results, policies in countries around the world, and the impact of climate change itself.

Determining potential emission reductions, for example, requires a baseline from which to measure those reductions; that is, an estimate of what forest-based emissions would be in the absence of the policy. Such baselines are uncertain because recent estimates of the contribution of forest-based emissions to global GHG releases range from 6 percent to 17 percent. Lower (or higher) baseline emissions would imply lesser (or greater) potential for reducing emissions from deforestation.

The costs of preserving forests as a way to mitigate climate change are also difficult to assess. They depend on the value of the land in alternative uses—that is, the opportunity costs, the benefits forgone when forests are left standing rather than being cleared for agricultural production or the harvesting of timber for wood products or for fuel. Opportunity costs change over time depending on economic, social, and institutional conditions such as population, technology, income, trade, and government policies. The models researchers use to estimate costs ideally would account for those factors over time, but assumptions about the future may plausibly differ and may have varying influences on cost estimates.

Estimates of the costs of forest-based mitigation also depend on the direct costs of putting preservation practices in place, such as those incurred in patrolling protected forests, and the costs of verifying the results of such practices. Verification includes determining whether the forest-based reductions in emissions of greenhouse gases would have occurred even in the absence of the incentives offered by the program, the extent to which the reductions are neutralized by program-related increases in emissions of greenhouse gases in other locations, and whether they are reliably measured over time. Although no standard definition of the costs of verification is widely accepted and no consensus has emerged on how to quantify or predict them, many experts believe that they are not high enough to negate the cost-effectiveness of some amount of forest-based mitigation.[15]

The cost-effectiveness of forest-based mitigation for the United States also depends on how other countries choose to use forests to achieve whatever goals they might have for reducing GHG emissions. Those choices could influence the amount of forest preservation that developing countries decide to undertake and how much developed countries are willing to pay to support that preservation—both of which will affect the cost of forest-based reductions in greenhouse gas emissions.

Finally, a changing climate itself would probably have an impact on the cost-effectiveness of forest-based mitigation. The location of forests, their species composition, and their productivity in storing carbon could be affected by changes in temperature, precipitation, and other factors that affect the health of forests. How the various effects of GHG emissions will affect forests over centuries is very uncertain. Forests could expand in some areas and contract in others, depending on the net effect of those influences. For instance, the growth rate of trees could increase as the amount of carbon dioxide in the atmosphere increases, although rates may eventually plateau as trees adjust to the new levels. However, existing forests may be threatened by changes in precipitation patterns. Those changing patterns and higher temperatures could also increase the number of forest fires. Similarly, diseases and pests that attack trees could become a greater threat in a warming climate.

CHALLENGES IN REDUCING FOREST-BASED EMISSIONS

Policies aiming to reduce global emissions of greenhouse gases by encouraging forest preservation in developing countries could impose costs or offer compensation as an incentive to reduce emissions. But, having other countries impose costs on developing countries to promote forest preservation would be very controversial. Forests are resources that many countries—including developed countries—have used in pursuing economic development. In addition, many developed countries have per capita GHG emissions that are higher than those of the developing countries that have large stores of carbon in their forests, suggesting that it would be seen as unfair treatment of nations that were poorer and had lower GHG emissions. Policies providing compensation to motivate efforts to preserve forestland would not raise such concerns, and the remainder of this section focuses on challenges involved in implementing such policies.

Three challenges would make it difficult to reduce forest-based emissions cost-effectively:

- Measuring changes in the amount of carbon stored in forests,
- Structuring incentives to reduce total forest-based emissions, and
- Improving governance in developing countries.

Measuring Changes in Carbon Storage

Forest-based mitigation programs need measurements of how the amount of carbon stored in forests in a country changes over time. Some experts conclude that advances in measurement techniques may reduce uncertainty enough within the next decade that such measurements would be useful for mitigation programs.[16]

Current Measurement Approaches

The amount of carbon stored in a forest depends on the product of four factors—the area, volume, and biomass of the forest, and the carbon content of the biomass:

Forest Carbon = Forest Area x Forest Volume x Forest Biomass x Carbon Content of Biomass.

Area is the amount of land covered by trees, and volume—the amount of wood—per unit of area depends on the size of the trees. The biomass—the living plant material produced from water and CO_2 by photosynthesis—per unit of volume depends on the density of the wood, which is influenced by factors such as a tree's location and species; carbon represents roughly half of that biomass.

Area and Volume of Forests

Data for estimating area and volume can come from on-the-ground inventories, in which trees in sample areas are counted and measured. The FAO compiles its Forest Resources Assessment every five years using forest inventories provided by individual countries. The quality of the country reports varies, however; not all countries identify forests using the FAO's definition.[17] Further, some countries may not have the resources needed to properly conduct inventories. Among the 50 nations with the greatest volume of wood in forests in 2005, 15 reported a loss in forest area between 1990 and 2005 but no accompanying change in volume per unit of area, suggesting that they may have simply assumed a constant volume over the 15-year period instead of reassessing that figure.[18]

Instruments mounted on airplanes or satellites also can provide data for estimating a forest's area and volume. Researchers can combine the data gathered by such remote-sensing activities with information from forest inventories to produce their estimates. Remote-sensing can cover a broader geographic scope and generate data more frequently than can an inventory. Standard remote-sensing technologies, which essentially offer a two-dimensional view of the forest from above, provide relatively good measures of forest area but not of forest volume. The U.S. Landsat program, started in 1972, is the longest continuously operating satellite system for observing the Earth. Landsat produces images using visible light and the heat radiated from objects on the Earth's surface. Using those data, researchers can estimate a forest's *area* with about 80 percent to 90 percent accuracy; in contrast, the accuracy of estimates of forest *volume* based on the data can be as low as 40 percent in forests with mixed species and overlapping tree canopies.[19] An important limitation of the data collected by Landsat is that the satellite's instruments cannot penetrate the clouds and smoke that often cover tropical rainforests.

More advanced and more expensive remote-sensing technologies can provide better measures of a forest's volume. Those instruments transmit pulses of radio waves (in the case of synthetic aperture radar, or SAR) or light from a laser (in the case of light detection and ranging, or LIDAR, technology) that can penetrate clouds, smoke, and the tree canopy to gather data on how objects on the Earth's surface reflect the emitted energy. SAR and LIDAR deliver a three-dimensional view of a forest and can be used to measure the elevation of the underlying land and the height of trees, allowing researchers to estimate the volume of forests with greater than 80 percent accuracy. Currently, only about nine SAR systems and one LIDAR system are operating on satellites, but that may change; among the current plans that countries around the world have for satellite missions between 2009 and 2016, 25 missions have SAR, two have LIDAR, and one has both.[20]

Biomass per Unit of Forest Volume

Data for estimating biomass per unit of volume come from statistical relationships that researchers have established by measuring, harvesting, drying, and weighing trees. The relationships used to estimate biomass in tropical forests—many of which contain 300 or more species of trees—are based on a limited sample of trees, and some researchers have suggested that additional sampling by region or ecological zone could improve current estimates of biomass. Nevertheless, species-specific relationships may not be needed; even in regions with highly diverse species, measuring the diameter of any type of tree at a height of about four and a half feet provides enough information about the tree's biomass content to explain more than 95 percent of the variation in biomass among different species.[21]

Carbon Content of Biomass

To impute the carbon content of biomass, researchers typically rely on broadly accepted values of about 50 percent. However, the carbon content of forest biomass can vary from 43 percent to 55 percent worldwide and from 43 percent to 49 percent in the tropics.[22]

Improving the Reliability of Measurements of Forest Carbon

Minimizing the uncertainty around each of the four factors used to estimate the amount of carbon stored in forests would be costly. Improving the precision of any one of the factors by a certain proportion would, however, have an equivalent impact on the precision of the final estimate, because the factors are combined multiplicatively to estimate the amount of forest carbon.[23] For example, correcting a 10 percent error in the carbon content of biomass would have the same impact as correcting a 10 percent error in forest area. Thus, information about the relative cost of equivalent percentage increases in the precision of different factors would help identify cost-effective improvements in measuring the amount of forest carbon. The need for such improvements could be assessed in light of the standards for accuracy set by any policies to reduce greenhouse gas emissions that might be put in place in the future. According to existing standards for the estimates of forest carbon that are included in voluntary reporting protocols and scientific inventories, those estimates should have no more than a 5 percent chance of overstating carbon storage.[24]

Improving the Capability of Developing Countries to Measure Changes in Forest Carbon

Even with techniques that could improve estimates of changes in forest carbon, developing countries may not have sufficient technical capabilities to make such estimates. However, those measurements would be needed to determine the compensation to be paid for preserving forestland. Those countries are unlikely to want outsiders in charge of assessing what they have accomplished with their sovereign resources for the purpose of determining the compensation due. Such concerns point to the importance of individual countries having the necessary personnel and technical capabilities to measure changes in their forest carbon rather than having to rely on outside groups to make those measurements.

For 13 of the developing countries that ranked in the top 25 nations for forest-based emissions from 1990 to 2005, some evidence is available about their ability to conduct forest inventories, their experience in acquiring and processing remote-sensing data, and their use of detailed measurements in reporting carbon stock (see Figure 5).[25] That evidence indicates that further capacity-building might improve countries' ability to produce useful measures of changes in forest carbon. Brazil, for example, might develop in-country expertise to conduct forest inventories that would complement the multiple inventories conducted by international consultants or donors. Nonetheless, Brazil is the only one of the 13 countries that has used advanced remote-sensing to monitor changes in forests and is one of only two that have performed their own evaluations of biomass and carbon content by forest type for purposes of carbon reporting (Mexico is the other one). Indonesia has produced only one forest inventory but has done so relying on in-country expertise. Although Indonesia has mapped forest area using standard remote-sensing technologies, unlike Brazil it has yet to use advanced remote-sensing to monitor changes in forests and has based its carbon reporting only on standard values for biomass and carbon content assigned by broad forest type.

Five of the 13 countries have no forest inventories and no experience in acquiring and processing remote-sensing data or in reporting forest carbon. [26]

Structuring Incentives to Reduce Forest-Based Emissions

Designing incentives to reduce worldwide forest-based emissions is particularly challenging because such incentives generate direct and indirect effects, often in opposite directions. The direct effect is the reduced rate of deforestation that is the initial response to the incentives. The indirect effects, commonly referred to as “leakage,” consist of responses in world markets to the direct effect and tend to offset the direct effect. For example, if incentives succeeded in reducing forest loss in countries receiving them, the resulting reduction in the supply of agricultural and timber products would probably lead to increases in their prices. Internationally, land values associated with those products would probably rise as a result. If that occurred, deforestation would probably increase in countries that were not receiving incentives, thereby reducing the cost-effectiveness of those incentives.

An effective incentive program requires projections of the amount of deforestation that would occur in the absence of the program and procedures that account for and minimize leakage. Various approaches to providing compensation have different implications both for the magnitude of leakage and for the extent to which programs end up paying for preservation in areas in which forests may not have been at risk of destruction. Thus, they can have an important influence on the cost-effectiveness of forest-based mitigation relative to other ways of reducing GHG emissions.

Baselines for Measuring Reductions

In general, policies that would provide compensation for forest preservation require national baselines from which to measure the reductions in GHG emissions associated with the policies. Consider, for example, a policy in which all developing countries would receive compensation for reducing their rates of deforestation. Under that policy, each country would have a baseline that reflects its historical rate of deforestation, which would be determined by the net rate observed over several past years because rates of deforestation vary significantly from year to year. Each country’s baseline would be used to indicate what its behavior would have been in the absence of the policy, including the net effects of activities that preserve forests and those that destroy them.

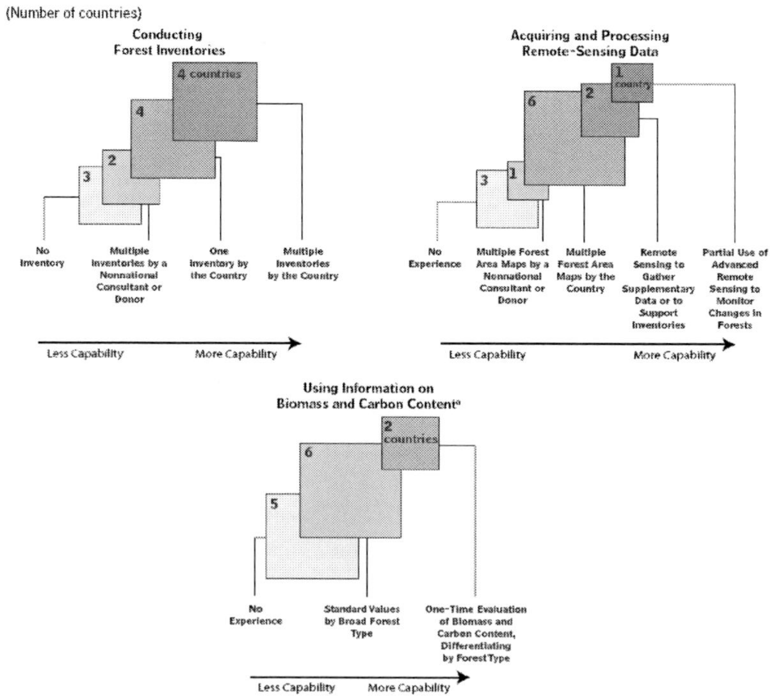

Source: Congressional Budget Office based on data from Pat Hardcastle and David Baird, *Capability and Cost Assessment of the Major Forest Nations to Measure and Monitor Their Forest Carbon* (Edinburgh: LTS International, April 7, 2008), www.ibcperu.org/doc/isis/ 11466.pdf.

Notes: Information on collecting data and measuring and reporting forest carbon is available for 13 of the 25 countries that made the largest contribution to forest-based emissions from 1990 to 2005 (see Figure 3). Those 13 countries are Bolivia, Brazil, Cambodia, Cameroon, China, the Democratic Republic of the Congo, Indonesia, Malaysia, Mexico, Myanmar, Papua New Guinea, Peru, and Venezuela. Canada, a developed country with vast forestland, provides a point of reference for the capability of developing countries to carry out those activities. Canada conducts multiple inventories on its own, uses advanced remote sensing, and, in reporting information on biomass and carbon content, differentiates by forest type and evaluates the same sample forest plots over time.

a. Obtaining the most useful information on biomass and carbon content requires differentiating by forest type and evaluating the same sample forest over time, but none of the 13 countries had that capability.

Figure 5. Capability of Developing Countries to Measure Changes in Forest Carbon.

Such a policy could have substantial direct effects in countries with high baseline rates of deforestation (such as Brazil and Indonesia), and those countries might reduce their rates of deforestation in response to the incentives offered under the policy. However, countries with low baseline rates (such as Gabon and Vietnam) would be likely places for leakage to occur because they would have little room for improvement and because world markets might offer them greater economic benefits from increasing deforestation than they would receive from reducing it.

Factors That Affect Leakage

In assessing incentives to reduce total forest-based emissions, it is important to consider several factors that influence leakage. First, less leakage would occur if countries representing a greater share of the world's forests participated in the compensation system.

Second, leakage would be smaller if more of the factors that led to deforestation in a country were local. For example, if a program led to the preservation of forests that would otherwise have been cleared to produce agricultural commodities for local markets, then the prices of goods in other countries would not increase. Without price increases, there would be no mechanism to prompt leakage.

Third, the willingness of people to change their patterns of consumption could also suppress leakage. If, for example, forest preservation caused supplies of certain locally produced goods to decline, people might purchase other goods instead—the production of which did not involve deforestation. The reduction in supply attributable to forest preservation would be accompanied by a reduction in demand, and prices would not change. Thus there would be no incentive to destroy forests elsewhere to replace the goods that had been locally produced before the forest preservation program began.

Finally, if production shifted to a location having a lower rate of greenhouse gas emissions, less leakage would occur. The CO_2e content per unit of output might be smaller in a new location if, for example, agricultural productivity per acre of land was higher there.

Assessing the Impact of Leakage

Leakage reduces the cost-effectiveness of programs aiming to achieve global net reductions in forest-based emissions.[27] When leakage occurs, global reductions in greenhouse gases fall short of the policy's intended reductions. In essence, then, leakage means that the actual cost of achieving a given reduction in global greenhouse gases is higher than it appears to be when measured within the limited geographic scope of the policy.

Anticipating the impact of leakage is difficult, however, because it cannot be easily measured. Measuring leakage requires identifying how reduced deforestation in one country affected markets for different products in all other locations. Researchers estimate leakage using economic data and modeling. To date, analyses of leakage from forest preservation have yielded estimates of emissions displaced to other locations that range from single-digit percentages to over 90 percent.[28]

Refining empirical estimates of leakage in the context of specific policies would improve the accuracy of assessments of the policies' costs. Estimates of the cost-effectiveness of forest-based mitigation, for example, are based on the assessment that achieving an additional reduction in GHG emissions through forest preservation is cheaper than doing so through an alternative strategy. For that assessment to hold, the direct cost advantage of reducing emissions through forest-based mitigation cannot be eliminated by leakage. Suppose, for example, that such a reduction suffered 50 percent leakage and that the reduction through the alternative strategy suffered none. For the forest-based mitigation to be worthwhile, its direct cost would have to be no more than half that of the alternative strategy. The greater the leakage differential, the greater the direct cost advantage needed to make forest-based mitigation cost-effective.

Structuring Incentives to Minimize Leakage

Funders of preservation programs might expand participation and minimize leakage by compensating developing countries that had been preserving forests over some period before the policy was in effect. For example, a country whose emissions from deforestation were less than half the global average could be compensated for maintaining that low rate. Alternatively, if those emissions were greater than or equal to the global average, a policy could compensate that country for lowering them. Because the factors affecting leakage vary across countries, however, policies that did not tailor incentives to the circumstances in each particular country might still have substantial leakage even if all countries had some incentives to reduce deforestation under the policy.

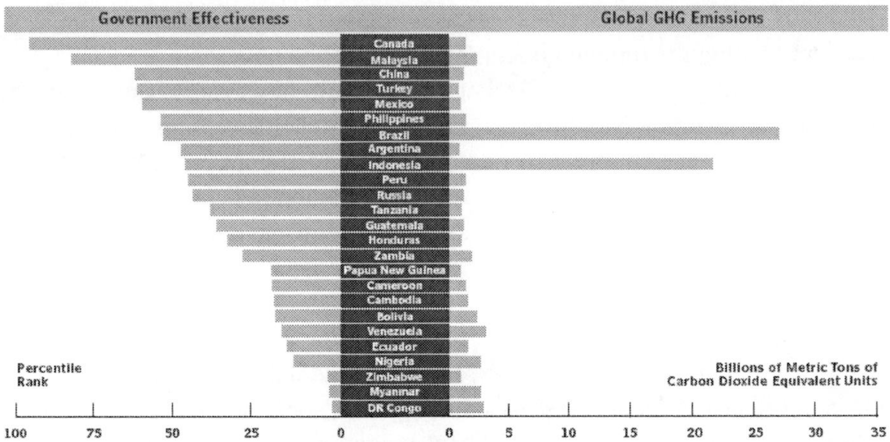

Sources: Congressional Budget Office based on data from World Bank,"Worldwide Governance Indicators" (2010), http://info .worldbank.org/governance/wgi/sc_ country.asp; and World Resources Institute, "Climate Analysis Indicators Tool (CAIT)" (2011), http://cait.wri.org.

Notes: "Government effectiveness" is a governance indicator reported by the World Bank. It measures the quality of public services, the quality of the civil service and the degree of its independence from political pressures, the quality of policy formulation and implementation, and the credibility of the government's commitments to such policies.

GHG = greenhouse gas; DR Congo = Democratic Republic of the Congo.

Figure 6. Government Effectiveness in Countries Responsible for 95 Percent of Global Forest-Based Emissions, 1990 to 2005.

Instead of basing incentives on differences in forest-based emissions relative to historical rates, a more complicated incentive formula could be used, incorporating projections about future pressures on forests arising from the policy itself, as well as from other economic, social, and institutional factors. Finding adequate data on the relevant economic, social, and institutional variables could be difficult, as could reaching agreement on which values for those variables—from among a range of defensible ones—to use in the projections.

Improving Governance in Developing Countries

The quality of a nation's governance influences its ability to implement policies and to deal with forces that undermine that ability, such as political instability and corruption. With inadequate governance, even programs that offer payment for reducing deforestation might not result in more preservation.

Governments' Effectiveness

One indicator of governance is government effectiveness. That indicator reflects a country's capacity for establishing reliable programs, such as those to reduce forest-based GHG emissions. According to the World Bank, which tracks and reports measures of governance, government effectiveness can be measured by the quality of public services, the quality of the civil service and the degree of its independence from political pressures, the quality of policy formulation and implementation, and the credibility of the government's commitments to such policies.[29]

Countries with an ineffective government might have difficulty producing verifiable reductions in emissions, even when programs for forest preservation are in place. Of the 25 countries with the largest forest-based emissions from 1990 to 2005, nearly three-quarters rank in the bottom half of all countries on measures of effective government (see Figure 6); they accounted for 55 percent of the world's forest-based GHG emissions. The two largest emitters, Brazil and Indonesia, are at roughly the 50th percentile for government effectiveness. By comparison, the government effectiveness rating for Canada, which is also in the top 25 producers of forest-based GHG emissions, is well above the 90th percentile.

In a weak government, the agencies responsible for preserving forests may have conflicting goals or inadequate authority for managing land use. For example, some countries may designate a forested area for both conservation and production—two competing uses of land.

Government agencies also may lack effective mechanisms for determining the amount of compensation and distributing it to indigenous peoples, local communities, and individuals who own forests, thereby undermining preservation efforts (see Box 2). In Brazil, for example, the process for determining compensation would be greatly improved if the appropriate agencies were able to map coordinates from satellite imagery to rural properties for which there is clearly established ownership. As of 2008, however, the government had only just started a pilot project to create that type of mapping and was doing so in only one of the nine Amazonian states.[30] In Cameroon, relatively strong forestry laws have been undermined because the local communities have not received adequate incentives from the national government to preserve forests.[31]

Poor governance may also mean that the rights to control forest resources are not clearly defined, making the gains from immediate use by clearing the land for agricultural production, timber, or wood for fuel more certain than the gains from preserving the forest. In Ghana, for example, landowners do not necessarily hold the right to the timber on their property. The lack of clarity about ownership creates incentives for farmers to log trees on their land to prevent timber companies with legal access to those trees from doing so and, in the process, damaging the farmers' cocoa crops.[32] In Brazil, a long-standing land statute, recently amended to counter abuse, allowed anyone who made "effective use" of the land to take a first step toward owning it, even without prior legal title. Forest clearing in that country, usually considered proof of such use, is a way to gain formal title to land owned by others and a way for landowners to prevent others from laying claim to their land.[33]

In some countries, the traditional claims of local populations that may support forest preservation often give way to statutory claims by others seeking to make greater use of forests' resources. In Indonesia, for example, one nongovernmental organization documented over 500 ongoing land conflicts between local communities and companies involved in producing palm oil.[34]

Corruption and Political Instability

Other aspects of governance, such as corruption and political instability, can also undermine efforts to preserve forests. National plans classifying forests for preservation might be intentionally structured to generate benefits for specific interests, but even if they are not, laws to protect forests can be ineffective if government officials ignore violations so they can personally benefit.

Little research has been done on the general relationship between corruption and forest loss. One study of African nations found no significant connection, and a broader study of countries across the globe found an association between a reduction in perceived corruption and lower rates of deforestation.[35] Some studies document how bribery undermines policies that protect forests, and anecdotal reports of corruption include cases in which timber has been harvested without a legal permit.[36] A recent report on forest reform in Liberia concluded that all existing logging contracts had been granted in violation of national laws; the report also noted widespread irregularities in the bidding process.[37] In the Democratic Republic of the Congo, some segments of the army, in addition to militia groups, profited from trade in illegally produced charcoal made from the forest resources of a national park.[38]

Box 2. The Role of Local Populations in Forest Preservation

Researchers emphasize the role of indigenous peoples and other forest-dependent communities in the success of efforts to preserve forests. An estimated 7 percent of public land in developing countries is reserved for those groups, who also own an estimated 57 percent of private lands in those countries. Overall, local populations control an estimated total of at least 19 percent of the land in developing countries.[1]

Local populations can contribute to the preservation of forests regardless of whether they have official control over that land. Forestland under national control may be difficult to preserve because of corruption or lack of resources for enforcement; engaging the participation of local communities might be a better way to preserve forests. For example, the Nepalese Forest Act of 1993 redirected the Department of Forests "from protecting national forests towards building forest user groups to manage all of the nation's forestland without interference from outside of the community."

One study of the impact of that law found that households' consumption of wood for fuel in areas in which user groups were involved in forest management was about 14 percent lower than in comparable areas without user groups, suggesting that harvests were being restrained to promote regeneration of degraded forests.[2] Other studies have found that territories officially recognized as belonging to indigenous peoples have been effective in reducing deforestation in the Brazilian Amazon.[3]

Conversely, preservation efforts may be undermined by populations that depend on forests for their livelihood. In the absence of incentives for preservation, community-controlled forestland might be converted to agriculture or managed for forest products in a way that increased greenhouse gas emissions. Forest preservation programs that fail to compensate local populations for new limits on their access to the resources they have traditionally relied on, or that fail to equitably distribute compensation for preservation, could jeopardize the support of local communities.

Local populations could oppose new forest preservation projects, and social unrest could undermine political support for such projects.

For example, the Tropical Forestry Action Plan, an international forest preservation initiative in the late 1980s and early 1990s, lost political support in part because the plan was designed without sufficient participation of indigenous peoples and other forest-dependent communities. It neglected differences between national concerns for forest preservation and such local concerns as community claims to forest resources, community benefits from logging, and assistance with economic development.[4] Recent international accords, such as the Cancun Agreement of the United Nations Framework Convention on Climate Change, assert the importance of involving forest-dependent communities in making decisions and managing forest preservation in a socially responsible way.

1. There is much more public land than private land in those countries. The Congressional Budget Office made these estimates using government data on land ownership in 18 developing countries that are among the world's 30 most heavily forested countries. The data underlying the estimates are from William D. Sunderlin, Jeffrey Hatcher, and Megan Lidle, From Exclusion to Ownership? Challenges and Opportunities in Advancing Forest Tenure Reform (Washington, D.C.: Rights and Resources Group, 2008), p. 8, Table 1, www.rightsandresources.org/documents/files/doc_736.pdf. The government data do not reflect the amount of forested land actively claimed by indigenous or other local communities.
2. Eric V. Edmonds, "Government Initiated Community Resource Management and Local Resource Extraction from Nepal's Forests," Journal of Development Economics, vol. 68, no. 1 (June 2002), pp. 89–115.
3. D. Nepstad and others, "Inhibition of Amazon Deforestation and Fire by Parks and Indigenous Lands," Conservation Biology, vol. 20, no. 1 (2006), pp. 65–73; and Britaldo Soares-Filho and others, "Role of Brazilian Amazon Protected Areas in Climate Change Mitigation," Proceedings of the National Academy of Sciences, vol. 107, no. 24 (2010), pp. 10821–10826.
4. Alexander Pfaff and others, Policy Impacts on Deforestation: Lessons Learned from Past Experiences to Inform New Initiatives (Durham, N.C.: Duke University, Nicholas Institute for Environmental Policy Solutions, June 2010).

Political instability also can undermine efforts to improve forest governance. Following an unconstitutional change in Madagascar's government in March 2009, for example, the United States and other countries suspended all aid to that country except humanitarian or emergency aid. Observers have noted recent reversals of the forest preservation in that country that had resulted from nearly two decades of involvement by donor countries in developing management systems for protected areas, promoting alternatives to slash-and-burn agriculture, and improving environmental practices.[39]

Efforts to Improve Governance

Countries that donate or lend money to developing countries have taken various approaches to improve the governance of forests. Those approaches include promoting sustainable forestry practices and standards for timber products, making the receipt of international lending contingent on reforms, funding the formulation of national action plans, and reducing the international debt of countries that establish conservation trust funds. But there are no quantitative evaluations of efforts to improve forest governance, and anecdotal evidence is mixed.

Programs to improve governance may fail because of insufficient funding or inadequate design.[40] In designing programs, for example, policymakers may not pay enough attention to the potential for leakage, which can undermine a program's goals. The European Union (EU) has been negotiating bilateral agreements with tropical countries. In cases in which negotiations have been completed, the EU has committed funding to strengthen enforcement of laws governing forests.[41] Then, following an agreed-upon period of investment, the EU will import designated wood products from those countries only if a certificate confirms the legality of their supply chain. However, some evidence indicates that timber from participating countries may be delivered to the EU through other countries, circumventing the aim of the agreements.

Policy Approaches for Reducing Forest-Based Emissions

Approaches that the United States and other developed countries might pursue to support forest-based mitigation in developing countries fall into two broad categories:

- Providing financial and technical assistance to governments interested in preserving forests and
- Creating demand in private markets for reductions in forest-based greenhouse gas emissions.

Providing financial and technical assistance would focus directly on addressing the challenges to reducing forest-based emissions. Creating demand in private markets could mobilize substantial funding for that effort. Although the two strategies can be pursued independently, they might work better in tandem. The viability of markets, for example, might depend on having in place a reliable program for achieving measurable reductions in forest-based emissions—the type of program that financial and technical assistance can help establish.

Assistance to Governments

Financial and technical assistance from developed countries could help address the more tractable challenges— measuring changes in forest-based greenhouse gases and creating incentives for cost-effective global net reductions in forest-based emissions. Whether such assistance could improve forest governance in developing countries is less clear.

Although the precise amount of assistance currently provided for forest-related activities is unknown, estimates to date suggest that providing sufficient funding to substantially reduce deforestation in developing countries could be difficult. All types of such assistance averaged roughly $560 million annually from the mid-1990s through 2008, or about $2.8 billion over a five-year period.[42] Another rough estimate puts the cost of setting up the capacity and infrastructure needed to operate a reliable program for reducing deforestation in a developing country that has the political will to carry out such a program at about $95 million over five years.[43] Together, the two estimates suggest that if historical amounts of aid were fully dedicated to preparing willing countries to run reliable programs for reducing deforestation, they might ready as many as 30 such countries to operate such programs.

Additional funds would be needed, however, to compensate those who reduce forest-based GHG emissions, and developed countries recognize that need.

Six nations— Australia, France, Japan, Norway, the United Kingdom, and the United States—recently pledged to provide $3.5 billion over three years to encourage forest preservation in developing countries. The pledge was in keeping with a collective commitment made by developed countries to provide an initial $30 billion in resources over the 2010– 2012 period to facilitate the adaptation to and mitigation of climate change. Although the commitments are explicit, whether they are in addition to existing ones is unclear, and whether the funds will be delivered is uncertain because some funds are contingent on the approval of the budgets of donor countries.[44] Fulfilling those commitments could ultimately depend on the extent to which forest preservation reduces the cost of meeting any goals that developed countries adopt for reducing emissions and on the competing demands in those countries for government funds.

With the challenges of improving governance and the potential limitations on funding, an effective strategy might be to direct assistance toward selected countries that have relatively strong governance and are rich in threatened forest resources. Of the 25 countries primarily responsible for global forest-based emissions in recent decades, Brazil, Indonesia, Venezuela, the Democratic Republic of the Congo, and Myanmar were the top five emitters, accounting for 69 percent of total forest-based emissions. Of the five, Brazil and Indonesia have comparatively strong governance (see Figure 6 on page 15).

They also have remaining forest resources that account for roughly 40 percent of the total forest area in tropical developing nations.[45] Establishing and supporting successful forest-based mitigation programs in those two countries could help preserve a significant proportion of tropical forests worldwide and might serve as a foundation for a similar effort encompassing other tropical developing nations. Such a broader effort would be important if leakage associated with the more focused program was significant. Because roughly 60 percent of the total forest area in developing tropical countries lies outside Brazil and Indonesia, significant leakage could occur, which would undermine the potential climate benefits of preserving the forests in those two countries.[46]

Markets for Reductions in Forest-Based GHG Emissions

Funding for forest preservation in developing countries might also be available in private markets. To date, markets for forest carbon have been largely voluntary rather than prompted by regulation. They exist primarily because of individuals' and businesses' perceptions of social responsibility and efforts to comply with expected climate policies even before the policies have been announced.

Historically, those factors have motivated over three-quarters of the transactions in the market for forest carbon. Markets directed only about $10.4 million in funding to forest preservation in 2009, avoiding 3.6 million tons of CO_2e emissions at an average price of $2.90 per ton.

They generated more funding for planting forests (about \$23.6 million at \$4.70 per ton) and for managing forests (about \$11.2 million at \$7.40 per ton)—activities that have less biological potential to reduce emissions than does preservation.[47] In 2010, encouraged by the attention paid to forest preservation in international policy discussions, market funding for preservation climbed to \$87.4 million, avoiding 19.4 million tons of CO_2e emissions at an average price of \$4.50 a ton. Funding for planting new forests declined slightly, to \$22.1 million, and funding for managing forests increased slightly, to \$13.4 million.[48]

To date, regulations have motivated only about one-quarter of the transactions in the markets for forest carbon. To increase market funding for preservation, forest-based reductions in GHG emissions could be sold in markets fostered by cap-and-trade programs or by tax credits that might accompany taxes on GHG emissions.[49] Such markets might motivate substantial increases in funding if forest-based mitigation abroad offered the opportunity to comply with required reductions in greenhouse gas emissions at a lower cost than that for reducing domestic emissions under a cap-and-trade program, for example. The more stringent the limit on emissions in a cap-and-trade program, the greater the potential for generating funding for forest preservation. Similarly, forest-based mitigation abroad might generate tax credits that could offset the impact of a carbon tax imposed on domestic GHG emissions, and the potential for generating funding for forest preservation would reflect the size of the tax.

Under either type of program, the potential funding for forest preservation would depend on the program's rules related to forest-based mitigation abroad. A cap-andtrade program, for example, might allow all or part of participants' GHG emission reductions to come from forest-based activities in foreign countries. It could also specify which countries were eligible for generating forest-based compliance. Similarly, a tax credit for forest preservation could fully or partially offset a domestic entity's related GHG tax liability, and the government could specify the countries in which forest-based activities would be eligible for generating credits.

The few existing regulatory programs do not credit emission reductions to forest preservation activities. However, some programs allow participants to meet a portion of their obligation by planting new forests. They include the Regional Greenhouse Gas Initiative (which requires power plants relying on fossil fuels in member states in the northeastern United States to reduce emissions) and the emissions trading systems of the European Union and New Zealand (which facilitate compliance with commitments to reduce emissions under the Kyoto Protocol of the United Nations Framework Convention on Climate Change).[50]

Even if markets established as a result of regulations provide greater financing for forest preservation, they are not necessarily well suited to support reliable programs for reducing forest-based greenhouse gas emissions. Meeting the three challenges—measuring changes in the amount of carbon stored in forests, creating incentives for cost-effective global reductions in forest-based emissions, and improving governance—generally requires up-front financing that is not directly linked to the purchase of the reductions in greenhouse gases accomplished through forest preservation. However, if regulatory programs specified the conditions under which they would credit emission reductions to forest preservation activities, the potential cost savings from use of those credits could motivate firms (or groups of firms in industries with high greenhouse gas emissions) to make investments to address the three challenges.

In particular, funding generated by a market-based program for reducing greenhouse gas emissions could help pay for improving measurement of forest carbon in developing countries or for preserving forests in countries where leakage might occur. The potential effect of such funding on governance in developing countries is less clear; it would depend, in part, on how governments would weigh the gains associated with qualifying for market participation against the costs they would incur in designing and implementing policies to preserve forests. For example, concerns that the distribution of associated benefits might create political tensions could make governments reluctant to increase forest preservation.

End Note for Summay

[1] Concerns about leakage are not unique to forest preservation programs. Leakage can also occur, for example, if sources subject to emission limits under national programs to lower greenhouse gas emissions relocate abroad to escape those limits, or if they lose market share to international competitors not subject to the limits.

End Notes

[1] Forests also influence climate by providing a cooling effect through evaporation and a warming effect because their relatively dark surfaces reflect less solar radiation from the Earth's surface.

[2] Adrian Deveny, Janet Nackoney, Nigel Purvis, and others, *Forest Carbon Index: The Geography of Forests in Climate Solutions* (Washington, D.C.: Resources for the Future and Climate Advisers, December 2009), Table A.6, p. 68, www.forestcarbonindex.org/ RFF-Rpt-FCI_small.pdf. The estimate does not include carbon stored below ground in tree roots and in soil. Soil alone accounts for 49 percent to 84 percent of all carbon stored in forests— varying with latitude, forest type, and soil type—but not all carbon in soil is emitted when forests are destroyed. Carbon from soil can account for 10 percent or more of CO_2 emissions from deforestation and degradation; soil-based emissions are particularly important in northern forests and peat swamps. See Matthew Fagan and Ruth DeFries, *Measurement and Monitoring of the World's Forests: A Review and Summary of Technical Capability, 2009–2015* (Washington, D.C.: Resources for the Future, December 2009), www.rff.org/Publications/Pages/ PublicationDetails.aspx?PublicatioID=20971; and Environmental Protection Agency, *Inventory of U.S. Greenhouse Gas Emissions and Sinks: 1990–2009*, USEPA 430-R-11-005 (April 2011), http://epa.gov/climatechange/emission s/usinventoryreport.html.

[3.] K.L. Denman and others, "Couplings Between Changes in the Climate System and Biogeochemistry," in S. Solomon and others, eds., *Climate Change 2007: The Physical Science Basis* (contribution of Working Group I to the Fourth Assessment Report of the Intergovernmental Panel on Climate Change, Cambridge University Press, 2007), www.ipcc.ch/publications_and_data/ar4/wg1/ en/ch7.html.

The impact of one ton of emissions of a greenhouse gas on climate differs depending on how that particular gas affects the balance within the Earth's atmosphere between incoming solar radiation and outgoing infrared radiation. To account for that difference, researchers report all GHG emissions in terms of carbon dioxide equivalent, or CO_2e. Because greenhouse gases vary in how long they stay in the atmosphere, that equivalence depends on the period over which the contribution of a metric ton is measured. By convention, CO_2e is usually measured over 100 years. A metric ton of CO_2e is the amount of a given greenhouse gas that makes the same contribution to global warming as a metric ton of CO2.

[4] G.R. van der Werf and others, "CO_2 Emissions from Forest Loss," *Nature Geoscience,* vol. 2 (November 2009), pp. 737–738.

[5] Ibid.

[6] S. Solomon and others, eds., *Climate Change 2007: The Physical Science Basis* (contribution of Working Group I to the Fourth Assessment Report of the Intergovernmental Panel on Climate Change, Cambridge University Press, 2007), www.ipcc.ch/ publications_and_data /publications_ipcc_fourth_assessment _report_wg1_report_the_physical_science_basis.htm.

[7] Hilda Munyua, *ICTs and Small-Scale Agriculture in Africa: A Scoping Study* (prepared for the International Development Research Centre, 2008), http://web.idrc.ca/uploads/user-S/ 12212542261Final_Report_HMunya.pdf.

[8] R. Bailis and others, "Mortality and Greenhouse Gas Impacts of Biomass and Petroleum Energy Futures in Africa," *Science,* vol. 308 (April 1, 2005), pp. 98–103.

[9] World Resources Institute, "Climate Analysis Indicators Tool (CAIT)" (2011), http://cait.wri.org.

[10] World Bank, World dataBank, "World Development Indicators & Global Development Finance," http://databank.worldbank.org/ ddp/home.do?Step=12&id=4&CNO=2.

[11] Lian Pin Koh and David S. Wilcove, "Is Oil Palm Agriculture Really Destroying Tropical Biodiversity?" *Conservation Letters*, vol. 1, no. 2 (June 2008), pp. 60–64, http://onlinelibrary .wiley.com/doi/10.1111/j.1755-263X.2008.00011.x/abstract.

[12] See Katherine Hamilton, Molly Peters-Stanley, and Thomas Marcello, *Building Bridges: State of the Voluntary Carbon Markets* (Washington, D.C.: Ecosystem Marketplace, 2008, 2009, 2010), www.ecosystemmarketplace.com. For data for 2007 to 2009, see Alexandre Kossoy and Philippe Ambrosi, *State and Trends of the Carbon Market 2010* (World Bank: Washington, D.C., May 2010), http://siteresources.worldbank.org/ INTCARBONFINANCE /Resources/State_and_Trends_of_the_Carbon_Market_2010_ lowjes.pdf.

[13] See Congressional Budget Office, *The Costs of Reducing Greenhouse Gas Emissions*, Issue Brief (November 2009).

[14] Environmental Protection Agency, Office of Atmospheric Programs, *EPA Analysis of the American Power Act in the 111th Congress* (June 14, 2010; revised June 30, 2010), www.epa.gov/ climatechange/economics/pdfs/EPA_APA_Analysis_6-14-10.pdf. In that accounting, the use of offsets from developing countries for compliance would decline by nearly 70 percent.

[15] For more information about the costs of verification, see Congressional Budget Office, *The Use of Offsets to Reduce Greenhouse Gases,* Issue Brief (August 2009).

[16] Molly Macauley and others, *Forest Measurement and Monitoring: Technical Capacity and "How Good Is Good Enough"* (Washington, D.C.: Resources for the Future, December 2009), www.rff.org/rff/ documents/rff-rpt-technical%20capacity _macauley%20et%20al. pdf.

[17] FAO defines a forest as "land spanning more than 0.5 hectares with trees higher than 5 meters and a canopy cover of more than 10 percent, or trees able to reach these thresholds (in place)." See Food and Agriculture Organization of the United Nations, *Global Forest Resources Assessment 2010* (Rome: FAO, October 2010), www.fao.org/forestry/fra /fra2010/en.

[18] P. Kauppi and others, "Returning Forests Analyzed with the Forest Identity," *Proceedings of the National Academy of Sciences*, vol. 103, no. 46 (November 14, 2006), pp. 17574–17579, http://phe.rockefeller.edu/docs/PNAS-Forests_final.pdf.

[19] See Fagan and DeFries, *Measurement and Monitoring of the World's Forests.* For forest area, "accuracy" refers to the percentage of pixels in the imagery that correctly identify the type of land cover; for forest volume, "accuracy" refers to the match between predictions from remote imagery and actual ground measurements.

[20] Ibid. Fagan and Defries point out that launches of planned satellite missions can fail or be delayed by budgetary and construction issues. Also, the operating life of satellites in orbit can be shorter than expected.

[21] Sandra Brown, "Measuring Carbon in Forests: Current Status and Future Challenges," *Environmental Pollution*, vol. 116, no. 3 (March 2002), pp. 363–372.

[22] Intergovernmental Panel on Climate Change, *2006 IPCC Guidelines for National Greenhouse Gas Inventories* (prepared by the National Greenhouse Gas Inventories Programme, edited by H.S. Eggleston and others, Hayama, Japan, 2006), Table 4.3, www.ipcc-nggip.iges.or.jp/public/2006gl/index.html; Fagan and DeFries, *Measurement and Monitoring of the World's Forests;* and Sandra Brown, "Measuring, Monitoring, and Verification of Carbon Benefits for Forest-Based Projects," *Philosophical Transactions of the Royal Society of London,* vol. 360 (2002), pp. 1669–1683.

[23] Paul E. Waggoner, *Forest Inventories: Discrepancies and Uncertainties*, RFF Discussion Paper 09-29 (Washington, D.C.: Resources for the Future, August 2009).

[24] Molly Macauley and others, *Forest Measurement and Monitoring: Technical Capacity and "How Good is Good Enough?"*

[25] Pat Hardcastle and David Baird, *Capability and Cost Assessment of the Major Forest Nations to Measure and Monitor Their Forest Carbon* (Edinburgh: LTS International, April 7, 2008), www.ibcperu.org/doc/isis/11466.pdf. The 13 countries are Bolivia, Brazil, Cambodia, Cameroon, China, the Democratic Republic of the Congo, Indonesia, Malaysia, Mexico, Myanmar, Papua New Guinea, Peru, and Venezuela.

[26] Canada, a developed country with vast forestland, provides a point of reference for the capability of developing countries to carry out those activities. Canada conducts multiple inventories on its own, uses advanced remote sensing, and, in reporting information on biomass and carbon content, differentiates by forest type and evaluates the same sample forest plots over time.

[27] Leakage is not specific to forest-based mitigation. It can occur with any policy for reducing greenhouse gas emissions that fails to address all potential emitters.

[28] Brian C. Murray, "Leakage from an Avoided Deforestation Compensation Policy: Concepts, Empirical Evidence, and Corrective Policy Options," in Charles Palmer and Stephanie Engel, eds., *Avoided Deforestation: Prospects for Mitigating Climate Change* (Oxford, United Kingdom: Routledge, 2009), pp. 151–172.

[29] World Bank, "Worldwide Governance Indicators" (2009),http://info.worldbank.org /governance/wgi/pdf/ge.pdf.

[30] Paulo Barreto, "Implications of the Climate Change Debate on Land Tenure in the Brazilian Amazon" (speech for the Conference on New Challenges for Land Policy and Administration, World Bank, Washington, D.C., February 14–15, 2008).

[31] Environmental Audit Committee, *Reducing Greenhouse Gas Emissions from Deforestation: No Hope Without Forests* (United Kingdom, House of Commons, June 16 2009), www.publications.parliament.uk/pa/cm200809/cmselect/ cmenvaud/30/3002.htm. See also Marcelin Tonye Mahop, "Forestry and Environmental Regulations, Legal Protection and Enforcement of Community Rights in Cameroon," *IUP Journal of Governance and Public Policy*, vol. 4, no. 1 (March 2009), pp. 7–21.

[32] C.P. Hansen and T. Treue, "Assessing Illegal Logging in Ghana," *International Forestry Review*, vol. 10, no. 4 (2008), pp. 573–590.

[33] Kathryn R. Kirby and others, "The Future of Deforestation in the Brazilian Amazon," *Future*, vol. 38 (2006), pp. 432–453.

[34] Sawit Watch, "Issues Surrounding Indonesian Palm Oil Industry" (memorandum submitted to the Delegation of the European Commission to Indonesia, July 24, 2009), http://normanjiwan.blogspot.com/2009_07_01_archive.html.

[35] Smith and others, "Governance and the Loss of Biodiversity," *Nature*, vol. 426, no. 6962 (November 6, 2003), pp. 67–70; and Edward B. Barbier, Richard Damania, and Daniel Léonard, "Corruption, Trade and Resource Conversion," *Journal of Environmental Economics and Management,* vol. 50, no. 2 (September 2005), pp. 276–299.

[36] Smith and others, "Governance and the Loss of Biodiversity"; Anne Casson and Krystof Obidzinski, "From New Order to Regional Autonomy: Shifting Dynamics of Illegal Logging in Kalimantan, Indonesia," in Luca Tacconi, ed., *Illegal Logging: Law Enforcement, Livelihoods and the Timber Trade* (London: Earthscan, 2007); and Douglas Dewitt Southgate Jr. and others, "Markets, Institutions, and Forestry: The Consequences of Timber Trade Liberalization in Ecuador," *World Development,* vol. 28, no. 11 (November 2000), pp. 2005–2012.

[37] Sustainable Development Institute, *Liberia—The Promise Betrayed* (January 2010), www.sdiliberia.org/sites/default/files/documents/ Promise%20Betrayed.pdf.

[38] United Press International, "Charcoal Trade Funds Congo Conflict," UPI.com (July 31, 2009), www.upi.com/Top_News/2009/07/31/Charcoal-trade-funds-Congo-conflict/UPI-8162124 9086595.

[39] Andrew C. Revkin, "Madagascar Forest Defenders Send S.O.S.," *New York Times*, March 20, 2009; and Linda Pressly, "Madagascar's Forests Plundered for Rare Rosewood," *BBC Radio 4, Crossing Continents* (August 5, 2010), www.bbc.co.uk/news/world-africa-10765418.

[40] For a review of design issues for efforts to conserve forests, see Alexander Pfaff and others, *Policy Impacts on Deforestation: Lessons Learned from Past Experiences to Inform New Initiatives* (Durham, N.C.: Duke University, Nicholas Institute for Environmental Policy Solutions, June 2010).

[41] To date, the EU has agreements with Cameroon, the Central African Republic, Ghana, Indonesia, Liberia, Malaysia, and the Republic of the Congo.

[42] Search of "Query Wizard for International Development Statistics" (Organisation for Economic Co-operation and Development, Paris), http://stats.oecd.org/qwids.

[43] That very rough estimate is based on the costs of past efforts to reform policies and institutions in developing countries. Whether the costs of previous interventions are a good indicator of the amounts needed for reform is not certain. In addition, the projects have not always achieved the desired outcomes. See Eliasch Review, *Climate Change: Financing Global Forests* (prepared by Johan Eliasch, 2008), p. 219, www.official-documents.gov.uk/ document/other/9780108507632/9780108507632.pdf; and personal communication to the Congressional Budget Office by a staff member of Chatham House, Royal Institute of International Affairs, United Kingdom, November 20, 2010.

[44] A recent agreement at the Cancun conference of the United Nations Framework Convention on Climate Change (UNFCCC) invited developed countries to submit an annual document presenting information on the resources they have provided to fulfill the commitment and on how developing countries obtain the resources. See UNFCCC, "Addendum, Part Two: Action Taken by the Conference of the Parties at Its Sixteenth Session," Decision 1/CP.16 (December 2010), p. 16, http://unfccc.int/resource/ docs/2010/cop16/eng/07a01.pdf#page=2.

[45] Hardcastle and Baird, *Capability and Cost Assessment of the Major Forest Nations to Measure and Monitor Their Forest Carbon,* www.ibcperu.org/doc/isis/11466.pdf.

[46] Brazil and Indonesia have slowed their deforestation rates in recent years. Estimated annual deforestation in Brazil declined from 2.9 million hectares during the 1990s to 2.6 million hectares during the subsequent decade. In Indonesia, it declined from 1.9 million hectares to 560,000 hectares. See Food and Agriculture Organization of the United Nations, *Global Forest Resources Assessment,* FAO Forestry Paper 163 (Rome: FAO, 2010), p. 21, www.fao.org/docrep/013/i1757e/i1757e00.htm.

[47] Katherine Hamilton and others, *Building Bridges: State of the Voluntary Carbon Markets 2010* (report by Ecosystem Marketplace, Washington, D.C., and Bloomberg New Energy Finance, New York, June 2010), http://moderncms.ecosystemmarketplace .com/repository /moderncms_documents/vcarbon_2010.2.pdf.

[48] David Diaz, Katherine Hamilton, and Evan Johnson, *State of the Forest Carbon Markets: From Canopy to Currency* (report by Ecosystem Marketplace, Washington D.C., September 2011), www.forest-trends.org/documents/files/doc_2963.pdf.

[49] In the United States, such provisions appeared in bills considered by the 111th Congress. State and regional programs in the United States may provide funding for reducing deforestation in developing countries. Under the Global Warming Solutions Act, for example, California

is actively considering such provisions. See Diaz, Hamilton, and Johnson, *State of the Forest Carbon Markets 2011,* www.forest-trends.org/documents/files/doc_2963.pdf.

[50] Under New Zealand's Emission Trading System, owners of older forests have to submit emission allowances for deforestation, but they are initially granted adequate allowances for that purpose. Owners may choose to sell those allowances if they do not plan to clear forests or if they anticipate that they can profit by selling the allowances and then—at a future time, when they clear forested land—by purchasing replacements at a lower cost.

In: Climate, Energy, and Environment. Vol. 1 ISBN: 978-1-62948-407-5
Editor: Jake S. Diaz

Chapter 2

OCEAN FERTILIZATION: THE POTENTIAL OF OCEAN FERTILIZATION FOR CLIMATE CHANGE MITIGATION*

National Oceanic and Atmospheric Administration

EXECUTIVE SUMMARY

Page 636 of House Report 111-366 that accompanies the Consolidated Appropriations Act of 2010 (Public Law 111-117) calls for the National Oceanic and Atmospheric Administration (NOAA) to "provide a report on the potential of ocean fertilization for climate change mitigation" to the House and Senate Committees on Appropriation within 60 days of enactment of the Act. Climate change mitigation includes any efforts to reduce climate change including reducing emissions of heat-trapping gases and particles, and increasing removal of heat-trapping gases from the atmosphere.

The oceans contain about 50 times as much carbon dioxide (CO_2) as the atmosphere, comprising around 38,118 billion metric tons of carbon compared to 762 billion metric tons in the atmosphere. What allows the oceans to store so much CO_2 is the fact that when CO_2 dissolves in surface seawater, it reacts with a vast reservoir of carbonate ions to form bicarbonate ions. This reaction effectively removes the dissolved gas form of CO_2 from the surface water, allowing the water to absorb more

* This is an edited, reformatted and augmented version of National Oceanic and Atmospheric Administration, dated 2010.

gas from the overlying air. This process, in combination with large-scale ocean circulation, has resulted in the transfer of between a quarter and a third of human-induced emissions of CO_2 from the atmosphere into the ocean since the beginning of the industrial revolution. Ocean biology enhances the ocean's ability to absorb CO_2 from the atmosphere as follows: plants in the ocean, mostly microscopic floating plants called phytoplankton, absorb CO_2 and nutrients when they grow, packaging them into organic material. Most of this organic material is consumed by other organisms and eventually recycled into CO_2 and nutrients near the surface. However, about 1-2% sinks into the deep ocean where it is converted back into CO_2 and nutrients by the action of bacteria. The resulting CO_2 and associated nutrients can remain out of contact with the atmosphere for hundreds of years, until ocean circulation brings the deep water to the surface and the cycle can begin again.

Recently, three U.S.-based companies have proposed fertilizing the ocean with iron and other essential nutrients to enhance the tendency of ocean biology to transfer CO_2 from the surface ocean to the deep ocean. These companies seek to deploy ocean fertilization as a climate change mitigation measure. In order for ocean fertilization to lead to climate change mitigation, three criteria must be met: (a) ocean fertilization must lead to increased growth of phytoplankton, packaging carbon and nutrients together into organic material; (b) this organic material must be transferred into the deep ocean so that it does not simply get recycled near the surface releasing its carbon back to the atmosphere; and (c) this transfer of carbon from the surface ocean to the deep ocean must result in a compensating transfer of carbon from the atmosphere into the surface ocean.

This report reviews two decades of ocean fertilization research that has focused primarily on adding the micronutrient iron to regions of the ocean that are iron-poor. A limited amount of research has focused on adding the nutrients nitrogen or phosphorus to areas that primarily lack those nutrients. However, since very small amounts of iron are required by phytoplankton to make organic matter (1 atom for every 50,000 to 200,000 atoms of carbon) compared to the nutrients nitrogen (1 atom for every 16 atoms of carbon) or phosphorus (1 atom for every 106 atoms of carbon), iron fertilization has been seen as the most feasible approach for altering the ocean's carbon cycle on a significant scale and has been the focus of scientific and commercial interest in the United States. This report focuses on the potential for fertilizing the ocean with iron to lead to climate change mitigation.

Ocean fertilization research has made an extremely valuable contribution to the scientific understanding of the ocean carbon cycle and its role in the global carbon cycle on time scales ranging from glacial episodes thousands of years in the Earth's past to today's changing climate. All reported ocean fertilization experiments have resulted in increased growth of phytoplankton (criterion (a)). However, the resulting transfer of organic material from the surface ocean into the deep ocean (criterion (b)) and the compensating transfer of carbon into the ocean from the atmosphere (criterion (c)) have not been verified. Thus, the research undertaken to date has not demonstrated that ocean fertilization is a mature technology for climate change mitigation. Modeling studies also suggest that the maximum possible impact of ocean fertilization on atmospheric CO_2 concentrations is likely to be a small fraction of what is needed to stabilize CO_2 concentrations at twice the preindustrial concentration, a general target for avoiding catastrophic climate change.

There are also environmental risks associated with ocean fertilization. Adding nutrients to the ocean causes changes in the structure and functioning of marine ecosystems, and these changes may result in unintended negative consequences. Scientists hypothesize that such consequences may include: decreases in productivity in ocean regions remote from the fertilization site; alteration of the relative abundance, size structures and diversity of higher trophic levels, including (but not limited to) economically important species; increases in deep ocean hypoxia or anoxia; increases in the oceanic production of the greenhouse gases nitrous oxide and methane; generation of toxin-producing algae; introduction of toxic chemicals at the fertilization site (depending on the composition of the fertilization material); and increases in ocean acidification with associated increased impacts in sub-surface ecosystems already vulnerable to ocean acidification. Concern about the environmental consequences of proposed large-scale ocean fertilization experiments has triggered the development of an international regulatory mechanism through the 1972 Convention on the Prevention of Marine Pollution by Dumping of Wastes and Other Matter, to which the United States is a Party.

NOAA executed an agreement with the National Academy of Sciences to establish a committee on America's Climate Choices, as required under the Department of Commerce Appropriations Act of 2008 (Public Law 110-161). This committee established four expert panels, including a Panel on Limiting the Magnitude of Future Climate Change. The Panel has described, analyzed, and assessed strategies for reducing the net future human influence on climate, including both technology and policy options. The Panel's report, released on May 19 2010, provided Congress with information on a wide range of climate mitigation options. The present report assesses the climate mitigation potential of a single proposed technology, ocean fertilization, as requested by Public Law 111-117.

I. INTRODUCTION

A. Definition of Climate Mitigation (Including Greenhouse Gasses)

The House Report (111-366, page 636) that accompanies the Consolidated Appropriations Act of 2010 (Public Law 111-117) calls for the National Oceanic and Atmospheric Administration (NOAA) to "provide a report on the potential of ocean fertilization for climate change mitigation" to the House and Senate Committees on Appropriation within 60 days of enactment of the Act. Climate change mitigation includes any efforts to reduce climate change including reducing emissions of heat-trapping gases and particles, and increasing removal of heat-trapping gases from the atmosphere. Key findings of the U.S. Global Change Research Program[1] and the Intergovernmental Panel on Climate Change[2] provide the motivation for exploring climate mitigation approaches and technologies.

Observations show that warming of the climate is unequivocal. The global warming observed over the past 50 years is due primarily to human-induced emissions of heat-trapping gases. These emissions come mainly from the burning of fossil fuels (coal, oil, and gas), with additional contributions from the clearing of forests and agricultural activities. Heat-trapping gases include carbon dioxide, methane and nitrous oxide.

Climate-related changes have already been observed globally and in the United States. These include increases in air and water temperatures, reduced frost days, increased frequency and intensity of heavy downpours, a rise in sea level, and reduced snow cover, glaciers, permafrost, and sea ice. A longer ice-free period on lakes and rivers, lengthening of the growing season, and increased water vapor in the atmosphere have also been observed. Over the past 30 years, temperatures have risen faster in winter than in any other season, with average winter temperatures in the Midwest and northern Great Plains increasing more than 7°F. Some of the changes have been faster than previous assessments had suggested.

Warming over this century is projected to be considerably greater than over the last century. The global average temperature since 1900 has risen by about 1.5°F. By 2100, it is projected to rise another 2 to 10°F. The U.S. average temperature has risen by a comparable amount and is very likely to rise more than the global average over this century, with some variation from place to place. Several factors will determine future temperature increases.

Increases at the lower end of this range are more likely if global heat-trapping gas emissions are cut substantially and/or technologies are employed to remove heat-trapping gases from the atmosphere and store them in places that can remain out of contact with the atmosphere for centuries to millennia.

NOAA executed an agreement with the National Academy of Sciences to establish a committee on America's Climate Choices, as required under the Department of Commerce Appropriations Act of 2008 (Public Law 110-161). This committee established four expert panels, including a Panel on Limiting the Magnitude of Future Climate Change. The Panel has described, analyzed, and assessed strategies for reducing the net future human influence on climate, including both technology and policy options. The Panel's report, released on May 19 2010, provided Congress with information on a wide range of climate mitigation options. The present report assesses the climate mitigation potential of a single proposed technology, ocean fertilization, as requested by Public Law 111-117.

B. How The Ocean Biosphere Stores Carbon Dioxide

The oceans store a vast amount of carbon dioxide (CO_2) in the forms of dissolved gas, carbonic acid, and carbonate and bicarbonate ions. The oceans currently contain about 50 times as much carbon dioxide as the atmosphere (around 38,118 GtC compared to 762 GtC in the atmosphere[3,4]), and are estimated to have absorbed between a quarter and a third of human-induced emissions of CO_2 since the beginning of the industrial revolution.[5] Of the 9.9 Gt of estimated global carbon emissions from fossil fuel burning and land use activities in 2008, 2.3±0.4 were absorbed by the ocean.[6]

The surface ocean absorbs CO_2 from the overlying air whenever its relative concentration (called its partial pressure) is higher in the air than it is in surface ocean water. Three factors make the ocean very effective at absorbing CO_2 from the atmosphere. The first factor is ocean chemistry. When CO_2 dissolves in surface sea water, it reacts with water and then with carbonate ions to form bicarbonate ions. These reactions effectively remove dissolved CO_2, lowering the partial pressure of CO_2 in the surface water and allowing the water to absorb more CO_2 from the overlying air. This process comes with a harmful side-effect, however, as the initial reaction with water forms a weak acid (carbonic acid), which reacts with carbonate ions thereby reducing carbonate ion concentrations. This decreased carbonate ion concentration makes those ions less available to the many marine organisms (such as corals and mollusks) that build their shells or skeletons out of calcium carbonate. This phenomenon is commonly referred to as ocean acidification. This effect of CO_2-linked ocean acidification may have deleterious impacts on marine ecosystems and their associated resources. NOAA scientists and their academic colleagues first quantified the impact of anthropogenic CO_2 on the calcium carbonate system in the global oceans in 2004[7]. Following the passage of the *Federal Ocean Acidification Research and Monitoring Act of 2009*, NOAA is establishing an ocean acidification program within the agency to conduct research and monitoring of ocean acidification and its impacts.

The second factor is large-scale ocean circulation. Seawater in major surface currents such as the Gulf Stream travels from the warm tropics to the cold Polar Regions, cooling as it goes. Carbon dioxide dissolves more readily in cold water than in warm water, so as the water cools, it accumulates CO_2. Eventually the water becomes cold and dense[8] enough to sink to the bottom of the ocean, carrying accumulated CO_2 into the deep ocean circulation system, where it can remain out of contact with the atmosphere for hundreds of years. Much of this carbon-rich deep water returns to the surface in the Southern Ocean, where strong winds and weak stratification[9] allow surface water and deep water to mix. Some remains in deep ocean currents, with a small fraction finally returning to the surface in the North Pacific, about 1600 years after it last had contact with the air in the North Atlantic. This ocean circulation system transfers CO_2 at today's concentrations into the deep ocean, removing it from contact with the atmosphere, while bringing waters to the surface that last had contact with the much lower CO_2 concentrations of the pre-industrial atmosphere. This exchange of old waters for new allows the entire ocean to participate in the chemical equilibration process described in the last paragraph, albeit on time scales of hundreds to thousands of years. NOAA scientists and their academic colleagues have traced the penetration of anthropogenic CO_2 (the fraction of atmospheric CO_2 that is associated with human activities such as burning fossil fuels and changing land use) into the deep waters of the Atlantic, and the intermediate waters of the Southern Ocean.[10] Oceanographers refer to the combined effect of ocean chemistry and ocean physics to transfer CO_2 from surface water to deep water as the "solubility pump."[11]

The third factor that makes the oceans so effective at absorbing CO_2 from the atmosphere is ocean biology. Just like trees do on land, plants in the ocean absorb CO_2 when they grow. In the open ocean, most of the plant growth is by microscopic floating plants called phytoplankton. Phytoplankton need CO_2, nutrients and sunlight to grow.

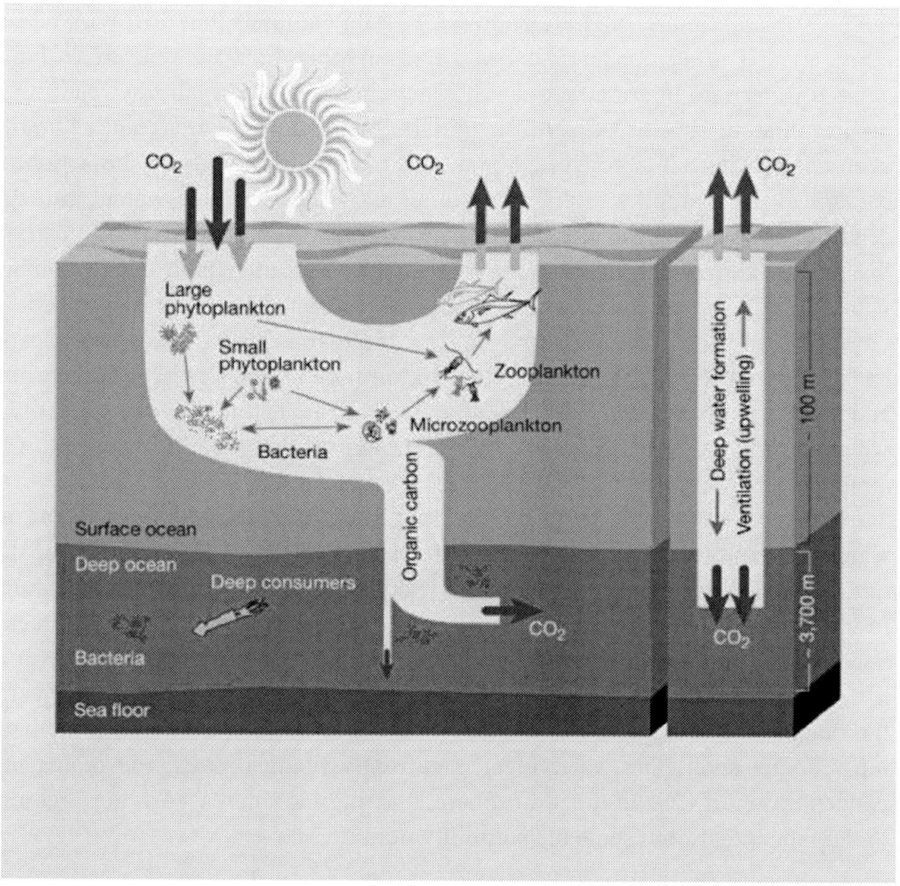

Figure 1. The ocean's biological pump (left) and solubility pump (right). Together these act to pump CO2 from surface water into the deep oceans, where it can remain out of contact with the atmosphere for hundreds of years. The solubility pump is driven by ocean physics and chemistry, whereas the biological pump is driven by ocean biology. Phytoplankton need CO2, nutrients and sunlight to grow. They convert CO2 and nutrients into organic matter, packaging carbon and nutrients together and forming the basis for marine food webs. Most of this organic matter is converted back to CO2 and nutrients near the surface when the phytoplankton are eaten by zooplankton (tiny animals that drift with ocean currents) and other consumers. But approximately 1-2% sinks into the deep ocean where it is converted back into CO2 and nutrients by the action of bacteria. Figure from Chisholm, S.W. 2000. Oceanography: Stirring times in the Southern Ocean. *Nature* 407: 685-687 doi:10.1038/35037696.

They convert CO_2 and nutrients into organic matter, packaging carbon and nutrients together and forming the basis for marine food webs. Most of this organic matter is converted back to CO_2 and nutrients near the surface when the phytoplankton are eaten by zooplankton (tiny animals that drift with ocean currents) and other consumers. But approximately 15-20% sinks out of the surface water into intermediate-depth waters below and a small fraction of that (approximately 10% or 1-2% of the initial biological production) sinks into the deep ocean where it is converted back into CO_2 and nutrients by the action of bacteria.[12] This process effectively pumps CO_2 from surface waters into the deep ocean. Oceanographers call this process the "biological pump." Carbon that is transferred into deep water by this process can remain out of contact with the atmosphere for hundreds of years. When deep waters enriched in nutrients and carbon return to the surface, the carbon stored there will escape back to the atmosphere unless phytoplankton take up the nutrients and carbon and begin the cycle again. Regions of unused nutrients thus represent locations where carbon stored in the ocean by the biological pump can leak back to the atmosphere.[13]

A useful analogy is to visualize the carbon stored in the deep ocean as water in a bath tub. The ocean carbonate system regulates the total size of the tub. The solubility pump and the biological pump are the faucets, adding water to the bath tub. The Southern Ocean, the Equatorial and North Pacific Ocean and other regions where deep ocean water rich in nutrients and carbon returns to the surface act to drain the bath tub. Over time scales of thousands of years, the faucets and the drains will balance each other, so that the water in the bath tub remains level. The water level is currently rising because the increasing concentration of CO_2 in the atmosphere is acting like increased water pressure in the faucet that represents the solubility pump.

C. Why the Ocean Biosphere Is Inefficient at Keeping Carbon Dioxide out of the Atmosphere (Nutrient Limitation, Light Limitation)

Vast regions of the world's oceans appear to contain all the necessary ingredients for abundant plant life (light, carbon dioxide, and the essential nutrients phosphorus, nitrogen and silicon), and yet support only sparse populations of phytoplankton. Oceanographers call these regions High Nutrient Low Chlorophyll (HNLC) regions, and they can be found in the North Pacific, the Equatorial Pacific and the Southern Ocean. As noted above, these regions are places where carbon stored by biological cycling can escape from the ocean. Scientists on the 1925-1927 Discovery Expedition observed HNLC conditions in the Southern Ocean and wondered if "minute quantities of inorganic compounds, as iron or manganese, or of organic compounds derived from the land" might be necessary for abundant phytoplankton growth.[14]

The late American oceanographer, John Martin, hypothesized that the missing ingredient is iron. Iron is an essential component in many of the enzymes and pigments necessary for plant growth. While it is the fourth most abundant element in the Earth's crust, iron in the ocean is either rapidly removed from the surface ocean by sticking to sinking particles or binds to chemicals that make it difficult for most plankton to use.[15] Martin's "iron hypothesis" is that phytoplankton in the HNLC regions suffer from iron deficiency, and this deficiency limits how much carbon those phytoplankton transfer from the surface ocean to the deep ocean via the biological pump.

This hypothesis is supported by work (including seminal research by NOAA scientists) showing that the optimal iron concentration for laboratory cultures of phytoplankton is higher than concentrations usually found in the open ocean.[16] Because plankton require very little iron (phytoplankton cells with sufficient iron generally contain 50,000-200,000 atoms of carbon for every atom of iron)[17] a few hundred thousand tons of soluble iron supplied by dust and sediments are all that is required to support the approximately 10 billion tons of organic carbon leaving the sunlit waters of the upper ocean each year.[18,19] Were Martin's hypothesis correct, it would imply that relatively small quantities of iron would be required to enable phytoplankton to make full use of the surface nutrients in the HNLC regions, greatly enhancing the ocean's biological pump and slowing the escape of dissolved CO_2 into the atmosphere that occurs in these regions.

The fact that iron can limit growth does not, however, mean that it is the only or even the dominant factor preventing phytoplankton from growing and "pumping" carbon and nutrients into the deep ocean. Competing hypotheses are that phytoplankton growth is held in check by a lack of light, particularly in the Southern Ocean where strong winds can stir surface water to great depths, plunging phytoplankton into darkness,[20] or that phytoplankton are quickly consumed by a highly efficient food chain of tiny zooplankton and microbes, which constantly recycle the phytoplankton back to their constituent nutrients.[21]

Martin's original statement of the "iron hypothesis" was an effort to explain the changes in the concentration of carbon dioxide in the Earth's atmosphere from glacial levels (around 200 ppm[22]) to the higher levels seen in warmer interglacial periods (around 280 ppm), a difference of around 170 GtC.

Previous work had suggested that an increase in the biological pump, particularly in the Southern Ocean, could account for this change.[23] While vast regions of today's oceans, including the Southern Ocean are iron-deficient, the iron hypothesis states that the much windier and dustier atmospheres in glacial times blew iron-rich dust onto the ocean, allowing phytoplankton to thrive and enabling a much larger transfer of carbon and nutrients into the deep ocean by a much more efficient biological pump. As some of this carbon dioxide comes out of the atmosphere, this could explain the large fluctuations in atmospheric carbon dioxide concentrations.[24] A competing hypothesis by J.R. Toggweiler is that the entire change in atmospheric CO_2 can be accounted for by a combination of three factors: an increase in the solubility of CO_2 in the colder surface waters of the glacial oceans; a decrease in the rate of return of deep water to the surface in the Southern Ocean due to changes in ocean circulation; and an increased capacity of the deep oceans to store CO_2 caused by adjustments in the calcium carbonate system.[25] To return to the bathtub analogy, Martin's hypothesis invokes an increase in the water being added to the tub by the biological pump faucet, whereas Toggweiler's hypothesis invokes the combination of an increase in the solubility pump faucet, a decrease in the Southern Ocean drain, and an increase in the size of the bathtub.

D. Proposals to Increase This Efficiency by Adding Macro/Micro Nutrients

Although Martin's "iron hypothesis" was motivated by his desire to explain glacial-tointerglacial changes in atmospheric CO_2 concentrations, in the paper describing the hypothesis, Martin advocated testing the idea by performing "realistic large-scale iron enrichment experiments" in iron-deficient regions of the modern ocean.[26] In a subsequent paper, Martin and co-workers made the link between the iron hypothesis and climate change mitigation: "oceanic iron fertilization aimed at the enhancement of phytoplankton production may turn out to be the most feasible method of stimulating the active removal of greenhouse gas CO_2 from the atmosphere, if the need arises."[27] Martin made his most famous statement of the iron hypothesis not in a scientific paper but in a July 1988 seminar at the Woods Hole Oceanographic Institution:

"I estimated that, with 300,000 tons of iron, the Southern Ocean phytoplankton could bloom and remove two billion tons of carbon dioxide.

Putting on my best Dr. Strangelove accent, I suggested that with half a ship load of iron... I could give you an ice age."[28]

Starting in 1993, an international community of oceanographers has tested the iron hypothesis in eleven open ocean iron addition experiments, and one experiment in which both iron and phosphorus were added to surface waters. The results of these experiments are summarized in Table 1 and are discussed in the next section. Other proposed ocean fertilization technologies include adding urea (a form of nitrogen) to nitrogen-starved areas of the ocean[29] and placing vertical pipes in the ocean to pump nutrient-rich deep water to the surface in areas that are devoid of all plant nutrients.[30]

Commercial interest in ocean fertilization technology began with the 1999 filing of US patent # 6,056,919 by Michael Markels, Jr., of Virginia-based GreenSea Venture. The patent is for a "Method of sequestering carbon dioxide" using iron and other fertilizers to enhance plant growth in the ocean. In the patent summary, Markels anticipated a market for carbon sequestration: "the governmental authority may allow credits on such balance sheet for the amount of carbon dioxide stated in such a report as being sequestered." In 2002, Russ George of the Planktos Foundation[31] fertilized a small patch of ocean near Hawaii[32] in anticipation of a voluntary market for the sale of carbon credits.

The other major venture in the area of iron fertilization is U.S.- based Climos, hoping to sell carbon credits on the regulated and unregulated emissions markets.[33] US-based Atmocean Inc. is designing pipes to pump cold, nutrient-rich deep water to the surface for carbon sequestration, fisheries enhancement and (by cooling the ocean surface) hurricane suppression.[34] Finally, Australia-based Ocean Nourishment Corporation seeks to commercialize urea fertilization (a form of nitrogen) for both carbon sequestration and fisheries enhancement.[35] Because far more nitrogen than iron is required for phytoplankton to make organic matter (roughly 1 atom of nitrogen for every 16 atoms of carbon, compared to 1 atom of iron for every 50,000 to 200,000 atoms of carbon), fertilizing the ocean with nitrogen is seen as far less feasible as compared to fertilizing the ocean with iron. Because of this, and because iron fertilization has been the most widely studied of the proposed ocean fertilization technologies, this report focuses on the potential for fertilizing the ocean with iron to lead to climate change mitigation.

Table 1. Summary of purposeful iron fertilization field experiments from: Strong, A.L., J. J. Cullen, and S.W. Chisholm. 2009. Ocean fertilization: Science, policy and commerce. *Oceanography Magazine* 22 (3):236- 261

Experiment	Year	Location	Duration	Magnitude	Rationale/Hypothesis Tested	General Conclusions
IronEx I Martin et al., 1994	1993	Eastern equatorial Pacific Ocean	10 days	450 kg Fe 64 km^2	• Iron limitation of productivity in HNLC region	• Iron limits phytoplankton growth rate, but patch subducted; broader implications of OIF unclear
IronEx II Coale et al., 1996	1995	Eastern equatorial Pacific Ocean	17 days	450 kg Fe 72 km^2	• Iron limitation of productivity in HNLC region	• Iron definitively limits productivity in equatorial Pacific. Larger bloom than IronEx I
SOIREE Boyd et al., 2000	1999	Southern Ocean-Australia; South of Antarctic Polar Front (APF)	13 days	1740 kg Fe 50 km^2	• Iron limitation of productivity in Southern Ocean, south of the Antarctic Polar Front (APF) • Fate of carbon fixed in bloom	• Iron limits productivity in Southern Ocean • No downward carbon transport observed
EisenEx Smetacek, 2001; Assmy et al., 2007	2000	Southern Ocean-Africa; in APF zone	21 days	4 tonnes $FeSO_4$ 38.5 km^2	• Iron limitation of productivity in Southern Ocean, along APF • Simulate Fe dust deposition to test whether Fe dust contributed to lower atmospheric CO_2 concentrations during glacial periods	• Iron limits productivity in Southern Ocean • Fate of the bloom uncertain • Iron only affected certain species of phytoplankton
SEEDS-I Tsuda et al., 2005	2001	Subarctic Pacific-Northwest	13 days	350 kg Fe 80 km^2	• Iron limitation of productivity in HNLC of subarctic Pacific • Fate of carbon fixed in bloom	• Iron limits productivity in subarctic Pacific • Floristic shift to diatoms • Downward carbon export minimal
SOFeX-N **SOFeX-S** Coale et al., 2004; Buesseler et al., 2004	2002	Southern Ocean-New Zealand; north and south of APF	30 days	N: 1712 kg Fe 225 km^2 S: 1260 Kg Fe 225 km^2	• Does OIF increase flux of carbon to deep ocean? • Silicate influence and geographic variability of response	• Increase in POC export flux, but magnitude is small relative to natural blooms
SERIES Boyd et al., 2004	2002	Subarctic Pacific-Gulf of Alaska	25 days	490 kg Fe 77 km^2	• Fate of carbon fixed in iron-induced bloom • Efficiency of carbon export to deep ocean	• Majority of carbon remineralized • Inefficient transport of carbon below thermocline
EIFEX Hoffmann et al., 2006; Jacquet et al., 2008	2004	Southern Ocean-Atlantic	35 days	7 tonnes $FeSO_4$ 150 km^2	• Iron addition impacts on phytoplankton community structure • Carbon sequestration efficiency and remineralization rates	• Shift away from picophytoplankton • Majority of carbon fixed was not remineralized • Unpublished "massive carbon export" paper by Smetacek et al.
FeeP Rees et al., 2007; Karl and Letelier, 2008	2004	Sub-tropical Northeast Atlantic-LNLC	21 days	5 tonnes $FeSO_4$ (+20 t PO_4) 25 km^2	• Interaction between iron and phosphorus controls on biological activity in the subtropical North Atlantic	• Increased N-fixation activity was observed • No increase in primary productivity • Carbon export not measured
SAGE Law et al., 2006	2004	Southern Ocean-250 km from New Zealand	15 days	5.4 tonnes $FeSO_4$ 100 km^2	• Iron addition's influence on sea-air gas exchange • CO_2 drawdown and dimethylsulfide (DMS) production	• Doubling of chlorophyll *a* but no significant DMS production and no significant CO_2 drawdown (preliminary results)
SEEDS-II Tsuda et al., 2007	2004	Subarctic Pacific-Northwest	26 days	491 kg Fe 64 km^2	• Monitor ultimate fate of bloom and carbon for longer time period than SEEDS-I	• No diatom bloom response • Increased zooplankton grazing
LOHAFEX NIO Press Release, 2009	2009	Southern Ocean-Atlantic	40 days	10 tonnes $FeSO_4$ 300 km^2	• Ecological shifts and fate of sinking carbon	• Increased zooplankton grazing • Negligible carbon export (preliminary results)

II. IRON FERTILIZATION EXPERIMENTS

The chemical and biological oceanography communities have been spurred by Martin's hypothesis to undertake a series of experiments where iron has been purposefully added to the ocean in various locations (Fig. 2, Table 1). Similar methods were used in most of the experiments: dissolved iron in the form of iron sulfate was mixed with seawater, an easy-tomeasure inert tracer was added to allow the scientists to keep track of the fertilized water,[36] and the mixture was pumped into the ocean from a moving ship. Ships then monitored the fertilized patch of seawater for 10 to 40 days, comparing chemical and biological parameters from the fertilized patch and unfertilized water outside of the patch. Additionally, a number of experiments have explored the impacts of natural addition of iron to the ocean associated with particular natural features, such as the plume of dust coming off the Sahara, and sources of iron in the sediments of the Kerguelen plateau in the Southern Ocean.

Because of the localized nature of these additions, it is possible to contrast regions that have received additional iron with adjacent regions that remain iron-poor, thus isolating the impact of iron fertilization. It should be emphasized that understanding the ocean's biological pump has not been the only motivation for these experiments: a significant motivation has also been to understand the role of iron in realistic ocean ecosystems. This is important as only a small fraction of ocean phytoplankton can be cultured in the laboratory, so it is by no means clear that responses found for individual species in the laboratory can represent responses of phytoplankton communities in nature. A list of purposeful ocean fertilization experiments is given in Table 1.

The questions examined in these field experiments include the following:

A. Does adding iron allow phytoplankton to grow more rapidly and make better use of available light and other nutrients?
B. Is this change associated with a change in the species of phytoplankton present at the fertilization site?
C. Does this additional growth result in a net uptake of nutrients and carbon?
D. Does the resulting "package" of carbon and nutrients (including iron) actually leave the surface ocean?
E. Does fertilization result in other chemical changes in or near the fertilized region?

Each of these questions is examined in the following subsections.

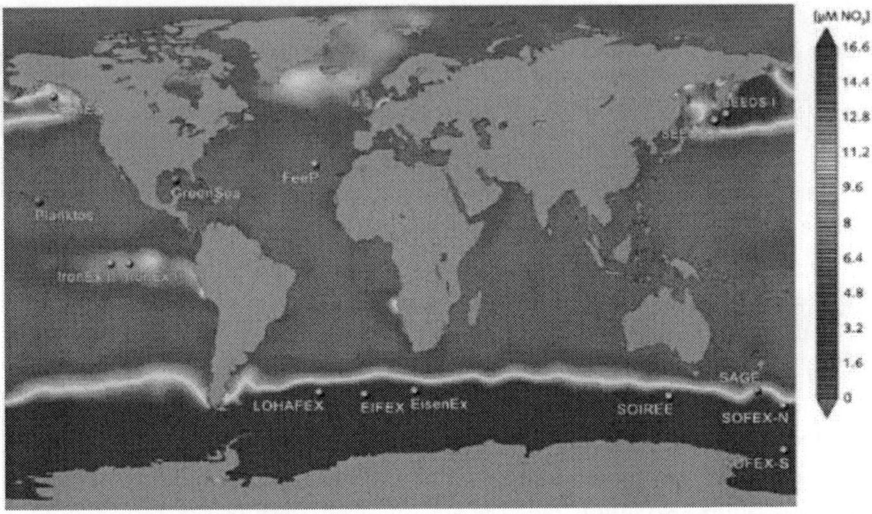

Figure 2. Locations of major iron fertilization experiments, including the pilot demonstrations of GreenSea Venture and Planktos. The color scale indicates the average annual surface nitrate concentration from the World Ocean Atlas 2005. Hotter colors indicate higher concentrations. The major HNLC regions in the Southern Ocean, the North Pacific and the Equatorial Pacific are clearly apparent as regions of high surface nitrate concentrations. Figure from Strong, A.L., J.J. Cullen and S.W. Chisholm, 2009. Ocean fertilization: science, policy and commerce. *Oceanography Magazine*, 22: 236-261.

A. Does Adding Iron Allow Phytoplankton to Grow More Rapidly?

The field experiments suggest that adding iron to HNLC regions of the ocean does make it easier for phytoplankton to grow. Increases in chlorophyll concentration in fertilized regions have been seen at all of the fertilization sites, indicating that iron does limit productivity in the equatorial Pacific, the Southern Ocean and the North Pacific. Measurements during a number of these experiments (including IronEx-II,[37] SOIREE,[38] and EisenEx[39]) demonstrated that this increase was due to an improvement of the ability of phytoplankton to use more of the available light. During the SoFEX experiment an increase in the inherent growth rate of large and small plankton was seen as well.[40] The results support the idea that adding iron causes an increase in the biomass of phytoplankton because it allows them to grow faster, rather than that iron acts to inhibit grazing in some way. Thus, the field experiments support the idea that, in general, increasing iron allows marine plants to take up more carbon. Researchers failed to observe a change in the total biological uptake of carbon in only one of the experiments (FeeP). However, in two others (IronEx I[41] and SoFEX[42]) local currents appear to have limited the duration and magnitude of the bloom.

B. Is Increased Growth of Phytoplankton Associated with a Change in the Species of Phytoplankton Present at the Fertilization Site?

The current conceptual picture of ocean ecosystems proposes that they function in one of two modes. In ecosystems in which one or more necessary nutrients (e.g., nitrogen, phosphorus or iron) are scarce, phytoplankton growth is limited. Small phytoplankton which are more efficient at taking up nutrients tend to dominate, but their populations remain low because they are eaten by small zooplankton. Small zooplankton can reproduce very rapidly whenever the phytoplankton population grows, keeping the phytoplankton population in check. These "nutrient-limited" ecosystems are thought to export relatively little carbon and nutrients to the deep ocean, with most material being recycled in the surface ocean. In contrast, when nutrients are abundant, larger phytoplankton tend to dominate the system. These larger phytoplankton are eaten by larger zooplankton, which take longer to reproduce and are less efficient at keeping the phytoplankton population in check. These ecosystems are considered more productive than nutrient-limited ecosystem. Inefficient grazing by large zooplankton allows some phytoplankton to sink, carrying their constituent carbon and nutrients into the deep ocean. In addition, the larger zooplankton and their predators (such as krill and small fish) tend to produce fecal pellets that sink rapidly, providing another means for pumping carbon and nutrients into the deep ocean.[43] Of particular interest in this respect are the large phytoplankton known as diatoms, which have for years been thought to play a disproportionately important role in highly productive marine ecosystems.[44] In this conceptual picture, adding iron to HNLC regions would be expected to change a nutrient-limited ecosystem into a more productive ecosystem, changing not only the amount of carbon taken up by phytoplankton, but the fraction of this carbon exported to the deep ocean.

The field experiments give moderate support to this picture. The IronEx II experiment saw a shift in species composition from small diatoms to large diatoms, with the genus *Pseudonitzschia* dominating.[45] The SOIREE experiment revealed a shift from small phytoplankton to diatoms after several days as did the EisenEx, SEEDS-1 and EIFEX experiments. However the SoFEX experiment saw all species increase in concentration and the SEEDS-II saw no diatom bloom response.

C. Does Additional Phytoplankton Growth Result in a Net Uptake of Nutrients and Carbon?

A larger uptake of carbon by phytoplankton does not necessarily imply a drawdown of surface nutrients- grazing of phytoplankton might simply recycle the organic matter back into inorganic form, thus failing to result in a net drawdown of nutrients and uptake of carbon. While the vast majority of iron fertilization experiments have shown clear drawdown of surface nutrients, the magnitude of this drawdown has varied greatly from experiment to experiment. For example, the SEEDS experiment in the North Pacific Ocean saw a drawdown of 15 micromole/kg, more than 80% of the initial nutrient concentration of 18 micromole/kg.[46] By contrast, during the SOFeX experiments[47] nitrate dropped by 2 micromole/kg within the fertilized patches, but this drawdown represented only about 10% of the initial nutrient concentration. Insofar as this experiment captured the full potential of iron fertilization in this region (something that is by no means clear) it would suggest that iron fertilization would not in fact keep carbon from escaping this region. The reasons for such differences are still not well understood, though the SOFeX investigators suggested that temperature and light limitation might play an important role.

D. Does the Resulting "Package" of Carbon and Nutrients (Including Iron) Actually Leave the Surface Ocean?

If iron fertilization is to serve as a climate mitigation technology, it must do more than stimulate productivity: the biological pump of carbon must be enhanced or the leak of carbon from the deep ocean must be plugged. It is at this point that the picture becomes less clear. Some reduction of the oceanic dissolved CO_2 concentrations has been seen in field experiments suggesting that the fertilized patch of ocean should have been taking up carbon from the atmosphere. Most of the purposeful fertilization experiments (e.g. Iron-Ex-II[48], SOIREE[49], and SEEDS[50]) found significant reductions (~20-60 ppm) in dissolved CO_2, though Iron-ExI specifically did not.[51] However, the time taken for the dissolved CO_2 in surface ocean water to come to equilibrium with CO_2 in the overlying air is very long (approximately 1 year),[52] so that the direct flux of carbon to the ocean that can be directly inferred from such experiments is actually very small.

An alternative to measuring the flux from the atmosphere to the surface ocean would be to measure the cause of such a flux, i.e. the export of carbon from the surface ocean to the deep ocean. However measuring this flux has not proved straightforward either. As seen in Table 1, in the SOIREE, SEEDS-1, SoFeX-N and SoFeX-S, SAGE and LoHAFEX experiments, the additional carbon export due to fertilization appears to have been very small.

Only the (as yet unpublished) EIFEX experiment saw a massive bloom of mat-forming diatoms that carried a significant amount of carbon to depth. A synthesis of iron fertilization experiments published in 2005[53] found that the ratio of carbon exported to iron added ranged from 1700 to 35000, significantly smaller than the much higher carbon-to-iron ratios found in ocean plankton. This suggests that much of the iron added during ocean fertilization experiments is simply lost from the system.

There are at least two possible explanations for the lack of observed carbon export. One is that the experiments were too small in scale and too short-lived. For example, during the SoFEX experiment, waters to which iron was added mixed with surrounding waters where iron and the concentration of organisms was much lower, leading to suggestions that this process prevented an intense bloom from developing.[54]

A second explanation, however, is that phytoplankton blooms do not necessarily lead to high carbon export. Recent observational work at Bermuda suggests that higher phytoplankton production may simply lead to more zooplankton consuming the phytoplankton higher in the water column.[55] Additionally, a recent global study comparing the amount of carbon-based particles collected in ocean sediment traps with satellite observations of phytoplankton growing near the sea surface revealed that during seasons with phytoplankton blooms, the fraction of carbon exported into deep waters is typically half that of the fraction exported during seasons when there are no phytoplankton blooms.[56]

E. Does Fertilization Result in Other Chemical Changes in or Near the Fertilized Region?

Ocean biology does more than just take up carbon. Diatoms in particular are thought to be sources of dimethylsulphide (DMS), an important gas that when released into the atmosphere, enhances cloud formation over the oceans. These clouds reflect sunlight and can exert a regional cooling effect. In addition, the cycling of organic matter back to its constituent carbon and nutrient can produce nitrous oxide and possibly methane as well, both of which are greenhouse gases with much higher heat trapping potential than CO_2.

Accordingly, a number of experiments have looked for such changes, with ambiguous results. For example, the SOIREE experiment found a tripling in DMS concentration over a two week period, but the SAGE experiment found no such change.

Certain species of diatoms are known to produce domoic acid, a neurotoxin associated with harmful algal blooms.[57] Iron enrichment experiments have generated large diatom blooms that in most cases are dominated by diatoms belonging to the genus Pseudo-nitzschia, including studies conducted in the northwest Pacific (SEEDS II),[58] northeast Pacfic (SERIES),[59] equatorial Pacific (IronEx II),[60,61] and the Southern Ocean (SOIREE,[62] EisenEx,[63], SoFeX,[64],[65] and Eifex[66])

The SERIES experiment demonstrated that while fertilization favored Pseudo-nitzschia, the phytoplankton genus that can produce domoic acid, detectable levels of this toxin were not found. However, a cruise in the same oceanographic area as SERIES (Ocean Station PAPA) has shown that oceanic Pseudo-nitzschia can produce measurable levels of toxin in its natural state and in response to iron addition experiments onboard ship and that domoic acid alters the phytoplankton community structure to benefit Pseudo-nitzschia.[67,68,69]

These results illustrate the still primitive state of understanding of how communities of phytoplankton actually behave and suggest that the consequences of ocean fertilization are still poorly known. Moreover, as some harmful algal blooms have been associated with the occurrence of excess nutrients in coastal waters,[70] the threat of harmful algal blooms remains a concern for certain applications of iron fertilization, particularly near the coast.

III. Models of Iron Fertilization

Evaluating the potential of iron fertilization for climate mitigation requires more than determining whether it can increase the biological pump of carbon and nutrients to the deep ocean. In fact a number of additional questions must be answered:

1. Can ocean fertilization take up enough carbon from the atmosphere have a mitigating effect on climate change?
2. Does exporting organic carbon to the deep ocean in a localized area lead to a net flux of carbon from the atmosphere to the ocean?
3. Can the flux of carbon from the atmosphere to the ocean be reliably quantified?
4. Are there undesirable consequences of fertilization?

All of these questions involve spatial scales that are much larger and time scales that are much longer than those associated with individual fertilization experiments. This means that the only way to address them is to use models to extrapolate from these specific results to the global scale. The following subsections introduce the models used to address these questions, and then treat each of them sequentially.

A. Modeling Ocean Circulation and Biology

The first computer models of the ocean circulation were developed by NOAA scientists during the late 1960s,[71] and similar models are now used by hundreds of oceanographers and climate scientists around the world. These models predict the circulation based on the conservation of heat, salt and momentum. Dividing the ocean up into millions of boxes, each of which contains some quantity of heat, salt, and momentum, an ocean model computes the forces that result from this distribution and uses these forces to generate a flow field. This flow field moves the heat, salt and momentum around, which then changes the driving forces of the flow. The current generation of ocean models are presently able to simulate the observed distributions of temperature and salt with reasonably good accuracy, so that the average temperature error over a box 100km on a side can be less than 1°C. They are also capable of simulating the distribution of trace gases, suggesting that they have relatively realistic rates of circulation.[72]

Models of ocean ecosystems and carbon cycling were embedded within these circulation models starting in the late 1980s.[73] The earliest models were extremely primitive. Seeking to reconstruct the patterns of biological activity that could produce the observed patterns of nutrients, they simply represented the effect of biology as taking up nutrients at some rate that depended on the local nutrient concentration, ignoring the details of how phytoplankton and their associated ecosystems actually respond to iron and iron cycling. Models that actually tried to predict plankton concentrations were included in the comprehensive physical models during the early 1990s[74] but did not in general include the impact of iron explicitly. It is only during the last decade that models with an explicit iron cycle that also resolve the impact of iron on different groups of plankton have emerged,[75] and these models still differ markedly from each other in terms of how they represent the impact of iron on ecosystems. These differences reflect significant scientific uncertainties as to how much soluble iron is delivered to the ocean, how it cycles chemically within the ocean, and what impact it has on changing the structure of ecosystems. Thus, while such models can be used to examine the interplay of different processes, their predictions are only as accurate as the biological and chemical dynamics that they encode. Researchers at a number of NOAA labs are trying to understand and model these fundamental dynamics which have implications for ecosystem structure and oceanic carbon uptake. However, it must be emphasized that many significant questions persist.

B. How Much Carbon Uptake Can Ocean Fertilization Produce, and How Rapidly?

As described above, the amount of carbon stored by the ocean depends both on how much carbon is injected into the deep ocean and how quickly it leaks out in HNLC regions. Insofar as iron fertilization acts to enhance the ocean's biological pump, it can increase the ocean carbon inventory only as fast as the natural rate of leakage, which in turn depends on the details of the ocean circulation. The first model study to examine this question with a comprehensive general circulation model was by Sarmiento and Orr,[76] using the Geophysical Fluid Dynamics Laboratory (GFDL) ocean general circulation model. This study assumed that iron fertilization would permit all nutrients brought to the surface to be used up, and then computed the impact on atmospheric CO_2. It was found that over 100 years, the ocean could take up a maximum of around 98-181 GtC , almost all of which was due to uptake in the

Southern Ocean, south of 30 degrees south latitude. The maximum decrease in atmospheric CO_2 concentration because of this carbon uptake was calculated to range from 46 to 85 ppm over the course of a century. This is smaller than the 100 ppm increase in CO_2 since the beginning of the Industrial Revolution. Another recent paper by Cao and Caldeira found that under a "business as usual" emissions scenario, a global scale drawdown of surface nutrients could reduce atmospheric CO_2 concentrations in the year 2100 from 965 ppm to 833 ppm.[77] Some of the difference between these two estimates arises from the fact that fertilization becomes more effective at removing carbon from the atmosphere as atmospheric CO_2 concentrations rise.

These model estimates of a potential atmospheric CO_2 reduction of up to 133 ppm represent an upper bound, unless major changes in ocean chemistry are allowed. More recent results[78] suggest that the region with the greatest leverage on atmospheric CO_2 concentrations is not the whole Southern Ocean, but rather the polar ocean around the Antarctic continent, because this is the region where unused nutrient levels are highest.[79] However, it is not clear how much of this leverage could in fact be realized, as these regions are dark during the wintertime and experience relatively low light levels even during the summer months. As a result, some models[80] show very little impact on atmospheric CO_2 from Southern Ocean fertilization. Other model studies that allow iron to change the ability of phytoplankton to make chlorophyll (the molecule responsible for harvesting light) have higher carbon uptake over a 100 year period. Aumont and Bopp[81] find a drawdown of 33 ppm over 100 years, though they note that in their simulations some nutrients remain at the surface throughout the year in the Southern Ocean. Using a similarly comprehensive biological model including an iron cycle, Sarmiento and co-workers[82] find an atmospheric drawdown of 35.3 ppm from Southern Ocean fertilization and 41.8 ppm from global fertilization over 100 years. It must be emphasized that these scenarios presuppose an infrastructure much more extensive than John Martin's "ship-load of iron;" they assume that millions of square kilometers of ocean are continually fertilized over hundreds of years, a feat that would be impractical if not impossible.

C. Local Export and Global Uptake

One aspect of ocean fertilization revealed by models is that there is a big difference between increasing the export of carbon from the ocean surface locally and actually storing more carbon in the ocean on a global scale. Just as stopping up one leak in a very leaky bathtub may simply divert the outflow to a different hole, fertilizing one part of the ocean and drawing carbon into it at that location may lead to an increase in the flux out of the ocean elsewhere. Negative feedbacks of this sort reduce the efficiency of iron fertilization as a climate mitigation mechanism.

One important negative feedback acts through the uptake of anthropogenic CO_2. Models predict that over time scales of thousands of years, the world's oceans will take up 80% of global CO_2 emissions from fossil fuel use and land use activities.[83] Viewed in this way, it becomes clear that fertilizing the ocean to take up carbon now will result in a trade off of reduced ocean carbon uptake in the future.

This effect is significant, compensating about 40% of the potential oceanic uptake of carbon over a course of a century.[84]

A second potential negative feedback arises because iron fertilization may simply concentrate production at one point in space and time rather than increasing it overall. If one imagines a row of faucets, all drawing from the same pipe, gushing into a bathtub, opening an additional faucet may simply cause the flow to redistribute itself amongst the faucets. This appears to be particularly important in tropical regions where the carbon escapes the ocean along the equator as nutrient- and carbon-rich waters are brought to the surface and warmed, but re-enter the ocean off-equator as biological cycling takes up these nutrients and with them, CO_2.

A third negative feedback involves changes in the ratio of carbon to nutrients within phytoplankton. This is based on the observation that Southern Ocean diatoms are actually less efficient at packaging carbon with other nutrients than are other plankton.[85] While this mechanism has been proposed in the scientific literature,[86] its magnitude has not been tested with realistic models.

A final negative feedback involves the cycling of nitrogen. As organic material sinks to depth and decomposes it consumes oxygen. If ocean fertilization is to sequester carbon in the deep ocean it must also reduce deep ocean oxygen concentrations. Such reductions, particularly on the eastern edge of the tropical oceans, can result in organisms using nitrogen rather than oxygen to consume organic matter, a process known as "denitrification." This process is also an important transient source of the greenhouse gas nitrous oxide. When waters that have lost their nitrogen come to the surface, the carbon in them is released back to the atmosphere. In the recent study by Sarmiento and co-workers, this process substantially reduced the efficiency of patch fertilization at a tropical site.

D. Can Oceanic Carbon Uptake due to Iron Fertilization Be Reliably Quantified?

In order for any climate change mitigation technology to gain global acceptance, it must be verifiable. This presents particular difficulties for ocean fertilization for three reasons. The first is that carbon does not stay in one place in the ocean, making it impossible to keep track of the carbon resulting from a particular fertilization exercise. The second is that the equilibration of carbon dioxide between the ocean and atmosphere takes many months, making it difficult to monitor the exchange of CO_2 between the atmosphere and the surface ocean.

The third is that any verification process must take into account the negative feedbacks outlined in the previous section, and these may be difficult to track and measure. As a result, verification cannot be accomplished by simply measuring carbon fluxes at the fertilization site. A recent model study[87] demonstrated that much of the uptake from the atmosphere from a localized fertilization takes place over time scales of many months and spatial scales of thousands of km, and amounts to a small change to a poorly known background flux of carbon between ocean and atmosphere. This paper also demonstrated that whether iron fertilization was an effective means of increasing efficiency of the biological pump depended on whether the iron remained associated with other nutrients and accessible to phytoplankton over time scales of decades. Again, verifying such continued association is not currently possible given the poor state of knowledge of the natural iron cycle.

E. Potential Negative Consequences

An alarming result from Sarmiento and Orr was the finding that fertilization in one region could cause productivity to collapse in another. In their model, fertilizing the Southern Ocean stripped out nutrients from waters that later moved northward, feeding tropical ecosystems. Further research[88] has suggested that the subpolar region of the Antarctic (north of about 60 degrees) is a region where fertilization would largely shift production from the tropics to the Southern Ocean while removing relatively little carbon from the atmosphere. Working with a more recent model, Aumont and Bopp [89] found that Southern Ocean biology could not, in fact, strip out all nutrients from Southern Ocean waters, and they also found that fertilization could reduce ocean productivity in regions at the edges of the subtropics. Moreover they found that if fertilization were turned off, the result would be that global biological productivity became lower than it would have been had no fertilization been applied.

An additional question addressed by model studies is whether iron fertilization might actually exacerbate the greenhouse warming problem. In particular, denitrification, noted above as a negative feedback on the biological pump, also produces the powerful greenhouse gas nitrous oxide. Jin and Gruber[90] suggested that such nitrous oxide production could offset a significant portion of the extra carbon taken up by the ocean.

Finally, a recent model study has quantified the impact of ocean fertilization on ocean acidification. Ocean fertilization relates to ocean acidification as follows: the goal of ocean fertilization is to cause a phytoplankton bloom in the surface ocean. The growing phytoplankton absorb dissolved carbon dioxide and transform it into organic matter via photosynthesis. This CO_2 absorption temporarily decreases surface ocean acidity, until carbon dioxide from the atmosphere replaces the absorbed gas. Meanwhile, some of the particulate organic carbon from photosynthesis (in the form of algal cells or particles of detritus) settles into deeper water where it is transformed back into CO_2 by the respiratory processes of marine animals and bacteria, thereby increasing the CO_2 concentrations in these deeper waters. Since most of this net release of CO_2 occurs in relatively shallow water directly below the sunlit surface layer,[91] the net result of ocean fertilization is to increase the amount of carbon dioxide, and hence acidity, in the shallow, subsurface ocean. This increased acidity changes the chemistry of ocean water with potential adverse effects on marine organisms, which depend on the existing chemistry for their growth and survival. For example, increased acidification of subsurface waters due to ocean fertilization will decrease the concentration of carbonate ions, which may have negative impacts on numerous marine organisms (such as corals and shellfish) that form their shells and skeletons out of calcium carbonate. The modeling study by Cao and Caldeira[92] found that massive global-scale ocean fertilization ameliorated surface ocean acidification to a minor degree (about 15% of the total impact needed to balance the increase in ocean acidification caused by increased atmospheric CO2 under a business-as-usual scenario) but at the cost of moving corrosive deep waters, capable of dissolving the shells of certain marine organisms, hundreds of meters closer to the surface.

IV. International Legal Framework and Applicable Law

A. Law of the Sea framework

Customary international law as reflected in the Law of the Sea (LOS) Convention provides the legal framework for all activities conducted at sea, including marine scientific research and ocean fertilization. Under LOS Article 238, "All States, irrespective of their geographical location, and competent international organizations have the right to conduct marine scientific research subject to the rights and duties of other States as provided for in this Convention." LOS Article 240 lays out the general principles for the conduct of marine scientific research including that "(d) marine scientific research shall be conducted in compliance with all relevant regulations adopted in conformity with this Convention including those for the protection and preservation of the marine environment."

B. Convention on the Prevention of Marine Pollution by Dumping of Wastes and Other Matter 1972 (London Convention)

In the case of ocean fertilization, the "Convention on the Prevention of Marine Pollution by Dumping of Wastes and Other Matter, 1972 (London Convention)" and the "1996 Protocol to the London Convention 1972 (London Protocol)" have emerged as the principal international regimes for addressing the potential impact of ocean fertilization on the marine environment. Although placing material in the ocean for ocean fertilization would not necessarily be considered ocean dumping, the 1972 London Convention and its 1996 Protocol require that placement of matter for a purpose other than disposal must not be contrary to the aims of the treaties, which focus on the prevention of pollution of the marine environment. Therefore, such placement into the water column should be subject to a system of review and evaluation to ensure the prevention of marine pollution. The United States is a Party to the London Convention (and has signed but not yet ratified the 1996 Protocol), and the domestic law that implements the London Convention in U.S. waters, and for activities taking place on U.S.-flagged vessels or involving material loaded in a U.S. port, is the Marine Protection, Research and Sanctuaries Act

(MPRSA). The US Environmental Protection Agency (EPA) is responsible for implementing the MPRSA.

In June 2007, in response to commercial interest in ocean fertilization and the potential of using ocean fertilization to generate carbon credits for sale on the regulated and voluntary emissions markets, the Scientific Groups of the London Convention and London Protocol issued a "Statement of Concern" on large-scale iron fertilization.[93]

This was endorsed by the Contracting Parties at their November 2007 meeting and expanded by the Scientific Groups in May 2008, with regard to large-scale and ocean fertilization activities other than just iron fertilization.[94] In October 2008, the Joint Meetings of Contracting Parties to the London Convention and Protocol adopted Resolution LC-LP.1 (2008) on the Regulation of Ocean Fertilization, which stated that "given the present state of knowledge, ocean fertilization activities other than legitimate scientific research should not be allowed."[95]

This resolution also stated that scientific research proposals should be assessed on a case-by-case basis using an assessment framework being developed by the London Convention and Protocol Scientific Groups.

As a Party to the London Convention, the U.S. supported Resolution LC-LP.1 (2008), agreeing that any ocean fertilization activity fit the definition of legitimate scientific research, as determined on a case by case basis using the assessment framework that is currently under development by the Scientific Groups of the London Convention and Protocol.

C. Antarctic Treaty System (ATS)

Six of the twelve open ocean iron fertilization experiments took place in the Southern Ocean, and the Southern Ocean remains the primary region of interest for ocean fertilization research. Much of the Southern Ocean falls under the geographical scope of the Antarctic Treaty and other agreements in the ATS, which cover all land and ocean space south of 60° South Latitude. Parties take several actions prior to conducting any activity in the Treaty area, which would include ocean fertilization experiments. The first action involves an environmental assessment process very similar to an Environmental Assessment or Environmental Impact Statement under the National Environmental Policy Act (NEPA). Parties also provide advance notification of the activity to other Treaty parties, and make a commitment to share data with other Treaty parties. Under the Protocol on Environmental Protection to the Antarctic Treaty, parties also have agreed to prevent marine pollution. As a party to the Antarctic Treaty, the U.S. may rely on this Treaty, the Protocol and the U.S. implementation authority to address ocean fertilization activities.

D. Magnuson-Stevens

One of NOAA's goals is to advance ecosystem approaches to fishery management. The Magnuson-Stevens Fishery Conservation and Management Act, 16 U.S.C. §§ 1801 et seq., as amended by the Magnuson-Stevens Fishery Conservation and Management Reauthorization Act (MSA) of 2006 (P.L. 109-479), acknowledges that regional fishery management councils have demonstrated significant progress in integrating ecosystem considerations in fisheries management. The Act regulates fishing in the U.S. Exclusive Economic Zone and covers a broad range of activities. For example, it would cover the catching, taking, or harvesting of all forms of marine animal and plant life with certain exceptions. The Act calls for NOAA to maintain a comprehensive program of research to support fishery conservation and management and to include research on, among other things, factors affecting the abundance and availability of fish (Sec. 404). In addition, fishery management plans must contain measures to protect, restore and promote the long-term health and stability of a fishery (Sec. 303). Plans must also describe and identify essential fish habitat (Sec. 303(a)(7)), and federal agencies must consult with NOAA on actions that are proposed, funded, authorized, or undertaken that may adversely affect the habitat (Sec. 305(b)). Ocean fertilization undertaken on a scale necessary to cause a measurable change in atmospheric concentration would likely result in large-scale changes in the structure and functioning of marine ecosystems. Proposed ocean fertilization activities would need to be carefully reviewed to see if they trigger essential fish habitat consultation or other MSA requirements. As noted in previous sections of this report, the science required to perform such a review is far from mature.

CONCLUSION

Fertilizing the ocean with iron (or other nutrients) has been proposed as a mechanism for mitigating climate change, by accelerating the uptake of CO_2

by the ocean. In order for such strategies to work, three criteria must be met: (a) ocean fertilization must lead to increased growth of phytoplankton, packaging carbon and nutrients together into organic material; (b) this organic material must be transferred into the deep ocean so that it does not simply get recycled near the surface releasing its carbon back to the atmosphere; and (c) this transfer of carbon from the surface ocean to the deep ocean must result in a compensating transfer of carbon from the atmosphere into the surface ocean. Research performed by NOAA, its academic partners and the wider research community has shown that while the first part of this chain of events is likely to occur (phytoplankton growth increases following ocean fertilization), the inevitability of the second two criteria is far from certain. Reliably quantifying any net uptake of carbon by the ocean following ocean fertilization is not currently possible, especially in the face of negative feedbacks that would tend to release carbon back to the atmosphere.

Research using ocean models to simulate the effects of ocean fertilization suggests that the maximum impact of ocean fertilization on ocean carbon uptake is likely to be a small fraction of what is required to stabilize atmospheric CO_2 concentrations at twice the preindustrial concentration. The maximum impact achieved in modeling studies requires continually fertilizing millions of square kilometers of ocean over hundreds of years, a feat that is not technologically feasible.

Ocean fertilization efforts that require manipulating ecosystems at a large scale could also potentially interfere with other legitimate uses of the sea. Potential negative consequences include undesirable changes in the structure and function of marine ecosystems, including ecosystems that support economically important fisheries; reductions in ocean productivity in regions affected by but remote from the fertilization site; increases in mid-ocean and deep ocean oxygen depletion; changes in ocean biogeochemistry that enhance the production of nitrous oxide and methane, greenhouse gases that have a higher heat-trapping potential than CO_2 on a molecule per molecule basis; stimulation of Harmful Algal Bloom forming species, and; net increases in ocean acidification.

End Notes

[1] Global Climate Change Impacts in the United States, Thomas R. Karl, Jerry M. Melillo, and Thomas C. Peterson, (eds.). Cambridge University Press, 2009.

[2] See, for example, IPCC, 2007: Climate Change 2007: Synthesis Report. Contribution of Working Groups I, II and III to the Fourth Assessment Report of the Intergovernmental

Panel on Climate Change [Core Writing Team, Pachauri, R.K and Reisinger, A. (eds.)]. IPCC, Geneva, Switzerland, 104 pp.

[3] 1 GtC (gigaton of carbon) is one billion metric tons of carbon, or 2,204.6 billion pounds of carbon. To put this into perspective, 1 Gt is equivalent to about 143 million African elephants or 247 thousand US Capitol domes.

[4] These numbers are from: Denman, K.L., *et a!*, 2007: Couplings between changes in the climate system and biogeochemistry. In: Climate Change 2007: The Physical Science Basis. Contribution of Working Group I to the Fourth Assessment Report of the Intergovernmental Panel on Climate Change [Solomon, S., D. Qin, M. Manning, Z. Chen, M. Marquis, K.B. Averyt, M.Tignor and H.L. Miller (eds.)]. Cambridge University Press, Cambridge, United Kingdom and New York, NY, USA.

[5] This estimate for the total transfer of CO2 into the ocean since the industrial revolution is derived from: Canadell, J.G., C. Le Quéréc, M.R. Raupach, C.B. Field, E.T. Buitenhuis, P. Ciais, T.J. Conway, N.P. Gillett, R.A Houghton, and G.Marland. 2007. Contributions to accelerating atmospheric CO2 growth from economic activity, carbon intensity, and efficiency of natural sinks. *Proc. Nat!. Acad. Sci. USA* 104, 18866-18870

[6] Le Quéré, C., M.R. Raupach, J.G. Canadell, G. Marland, L. Bopp, P. Ciais, T. J. Conway6, S.C. Doney, R.A. Feely, P. Foster, P. Friedlingstein, K. Gurney, R.A. Houghton, J. I. House, C. Huntingford, P. E. Levy, M.R. Lomas, J. Majkut, N. Metzl, J. P. Ometto, G.P. Peters, I.C. Prentice, J.T. Randerson, S.W. Running, J.L. Sarmiento, U. Schuster, S. Sitch, T. Takahashi, N. Viovy, G.R. van der Werf, and F.I. Woodward.*!*. 2009. Trends in the sources and sinks of carbon dioxide. *Nature Geoscience* 2: 831-836, DOI: 10.1038/NGEO689

[7] Feely, R.A., C.L. Sabine, K. Lee, W., Berelson, J. Kleypas, V.J. Fabry., and F.J. Millero. 2004. Impact of Anthropogenic CO2 on the CaCO3 System in the Oceans. Science 305 (5682), 362. [DOI: 10.1126/science.1097329]

[8] Seawater density is determined by salinity and temperature, with saltier water being denser than less salty water, and colder water being denser than warmer water. Rain falling on the ocean can make surface water less salty and less dense, whereas evaporation or sea ice formation (when sea ice forms, only water freezes and the salt is left behind) make water more salty and more dense.

[9] In most regions of the world's oceans, water is stratified with less dense water floating on top of water that is denser.

[10] Sabine, C.L., R.A. Feely, N. Gruber, R.M. Key, K.Lee, J.L. Bullister, R. Wanninkhof, C.S. Wong, D.W.R. Wallace, B. Tilbrook, F.J. Millero, T.-S. Peng,A. Kozyr, T. Ono, and A.F. Rios. 2004. The Oceanic Sink for Anthropogenic CO2. *Science* 305 (5682), 367. [DOI: 10.1126/science.1097403]

[11] Volk, T., and M.I. Hoffert (1985), Ocean carbon pumps: Analysis of relative strengths and efficiencies in ocean driven atmospheric CO2 changes, in *The Carbon Cycle and Atmospheric CO2: Natural Variations Archaean to Present*, Geophys. Monogr. Ser., vol. 32, edited by E. T. Sundquist and W. S. Broecker, pp. 99–110, AGU, Washington, D. C

[12] Dunne, J.P., J.L. Sarmiento, and A. Gnanadesikan, 2007: A synthesis of global particle export from the surface ocean and cycling through the ocean interior and on the seafloor. *Global Biogeochemical Cycles*, 21, GB4006, doi:10.1029/2006GB002907

[13] Watson, A.J. and J.C. Orr.2003. Carbon dioxide fluxes in the global ocean, in *Ocean Biogeochemistry: The role of the ocean carbon cycle in climate change*, Edited by M.J.R Fasham, Springer-Verlag, pp. 123-144.

[14] See comments on the 1925-1927 *Discovery* Expedition in Hardy, 1967, quoted by Martin, J.H. R.M. Gordon, 1988. Northeast Pacific iron distributions in relation to phytoplankton productivity. *Deep-Sea Research* 35:177-196

[15] Rue, E.L. and K.W. Bruland, Complexation of iron(III) by natural organic ligands in the Central North Pacific as determined by a new competitive ligand equilibration / adsorptive cathodic stripping voltametric method, *Marin. Chemistry,* 50, 117–138, 1995.

[16] Sunda, W.G. and S.A. Huntsman, 1997: Interrelated influence of iron, light, and cell size on marine phytoplankton growth, *Nature*, 390, 389-392.

[17] De Baar, H.J.W., L.J.A. Gerringa, P. Laan, and K.R. Timmermanns, 2008: Efficiency of carbon removal per added iron in ocean iron fertilization, *Mar. Ecol. Progr. Ser.*, 364. 269-282.

[18] Dunne, J.P., J.L. Sarmiento, and A. Gnanadesikan, 2007: A synthesis of global particle export from the surface ocean and cycling through the ocean interior and on the seafloor. *Global Biogeochemical Cycles*, 21, GB4006, doi:10.1029/2006GB002907

[19] Estimates of the delivery of iron to the ocean by atmospheric dust range between 10 and 35 million tons but only a small fraction of this iron is soluble and thus available to plankton. See: Jickells, T. and L. Spokes, 2001: Atmospheric iron inputs to the ocean, in *The Biogeochemistry of Iron in Seawater*, edited by D.R. Turner and K.A. Hunter, pp. 85-121, John Wiley, NY.

[20] Mitchell, B.G., E.A. Brody, O. Holm-Hansen, C. McClain, and J. Bishop. 1991. Light limitation of phytoplankton biomass and macronutrient utilization in the Southern Ocean. *Limnology and Oceanography*, 36: 1662-1677

[21] Frost, B.W. 1991. The role of grazing in nutrient-rich areas of the open sea. *Limnology and Oceanography*, 36: 1616-1630.

[22] Concentrations of C02 in the atmosphere are typically reported in parts per million by volume, or ppm for short. 200 ppm means that for every million volumes of air, 200 consist of C02. For updated values of atmospheric CO2 concentration, please see: http://www.esrl.noaa.gov /gmd/ccgg/trends/index.html#global

[23] Sarmiento J.L. and J.R. Toggweiler, 1984: A new model for the role of the oceans in determining atmospheric pCO2, *Nature* 308,621-624.

[24] Martin, J.H. 1990. Glacial-Interglacial CO2 Change: The Iron Hypothesis, *Paleoceanography*, 5(1), 1–13.

[25] Toggweiler, J.R. 1999. Variations of atmospheric CO2 by ventilation of the ocean's deepest water. Paleoceanography 14: 571-588

[26] Martin, J.H. 1990. Glacial-Interglacial CO2 Change: The Iron Hypothesis, *Paleoceanography*, 5(1), 1–13.

[27] Martin, J.H., R.M. Gordon, and S.E. Fitzwater. 1990. Iron in Antarctic waters. *Nature* 345: 156-158

[28] Martin, J.H. 1990. A new iron age, or a ferric fantasy. *U.S. JGOFS News* 1(4): 5, 11

[29] See Young E (2007) A drop in the ocean: Can dumping tonnes of fertiliser into the oceans really turn back the tide of global warming? *New Scientist*, September 15 2007, 42-45

[30] Lovelock J and C. Rapley. 2007. Ocean pipes could help the Earth to cure itself. *Nature* 449: 403. doi:10.1038/449403a

[31] Now Planktos Science: http://www.planktos-science.com. Website accessed on 2 February 2009.

[32] Schiermeier, Q. 2003. Climate change: The oresmen. *Nature* 421, 109-110 (9 January 2003) | doi:10.1038/421109a

[33] Whilden, K., M. Leinen, D. Whaley, and B. Grant. Ocean Fertilization as an Effective Tool for Climate Change Mitigation. in *IETA Greenhouse Gas Market Report 2007: Building Upon a Solid Foundation: The emergence of a global emissions trading system*. D. Lunsden, editor. International Emissions Trading Association, Geneva, Switzerland. 143 pp.

[34] Information from http://www.atmocean.com/index.htm, accessed on 2 February 2010. A report on an at-sea trial of the Atmocean pump technology is published as: White, A., K. Bjorkman, E. Grabowski, R. Letelier, S. Poulos, B. Watkins, and D. Karl. 2010. An open ocean trial of controlled upwelling using wave pump technology. *Journal of Atmospheric and Oceanic Technology*. 27: 385-396. doi: 10.1175/2009JTECHO679.1

[35] Glibert, P.M. et al (56 coauthors) 2008. Ocean urea fertilization for carbon credits poses high ecological risks. Marine Pollution Bulletin 56:1049–1056 doi:10.1016/j.marpolbul.2008.03.010

[36] In the report of IronEx I, the first open-ocean iron fertilization experiment, Martin and co-authors identified the development of the ability to measure ultra-trace concentrations of the inert gas sulfur hexafluoride, SF6, as the major technological breakthrough that made ocean fertilization experiments possible by enabling the researchers to mark a patch of sea water and track it for long periods of time. See Martin, J.H. *et al.* 1994. Testing the iron hypothesis in ecosystems of the equatorial Pacific Ocean. *Nature* 371: 123-129 doi:10.1038/371123a0

[37] Coale, K. H., K. S. Johnson, S. E. Fitzwater, R. M. Gordon, S. Tanner, F. P. Chavez, L. Ferioli, C. Sakamoto, P. Rogers, F. Millero, P. Steinberg, P. Nightingale, D. Cooper, W. P. Cochlan, M. R. Landry, J. Constantinou, G. Rollwagen, A. Trasvina and R. Kudela. 1996. The IronEx-II mesoscale experiment produces massive phytoplankton blooms in the equatorial Pacific. Nature, 383, 495-501.

[38] Boyd, P.W., A.J. Watson, C.S. Law, E.R. Abraham, T. Trull, R. Murdoch, D.C.E. Bakker, A.R. Bowie, K.O. Buesseler, H. Chang, M. Charette, P. Croot, K. Downing, R. Frew, M. Gall, M. Hadfield, J. Hall, M. Harvey, G. Jameson, J. LaRoche, M. Liddicoat, R. Ling, M.T. Maldonado, R.M. McKay, S. Nodder, S. Pickmere, R. Pridmore, S. Rintoul, K. Safi, P. Sutton, R. Strzepek, K. Tanneberger, S. Turner, A. Waite, J. Zeldis. 2000. A mesoscale phytoplankton bloom in the polar Southern Ocean stimulated by iron fertilization, *Nature*, 407, 695-702. doi:10.1038/35037500

[39] Gervais, F., U. Riebesell and M.Y. Gorbunov, 2002: Changes in primary productivity and chlorophyll *a* in response to iron fertilization in the Southern Polar Frontal Zone, *Limnol. Oceanogr*, 47, 1324-1335.

[40] Hiscock, M.R., V.P. Lance, A. Apprill, R.R. Bidigare, Z.I. Johnson, B.G. Mitchell, W.O. Smith and R.T. Barber, 2008: Photosynthetic maximum quantum yield increases are an essential component of the Southern Ocean phytoplankton response to iron, *Proc. Natl. Acad. Sci.*, 105, 4775-4780.

[41] Coale, K.H, K.S. Johnson, S.E. Fitzwater, S.P.G. Blain, T.P. Stanton and T.L. Coley, 1998: IronEx-I, An in-situ iron-enrichment experiment: Experimental design, implementation and results, *Deep Sea Res.,Part II*, 919-945.

[42] Krishnamurty, A., J.K. Moore and S.C. Doney, 2009: The effects of dilution and mixed layer depth on deliberate iron fertilization: 1-D simulations of the Southern Ocean Iron Experiment, (SOFeX), J. Marine Systems, 71, 112- 130.

[43] Dunne, J. P., J. L. Sarmiento, and A. Gnanadesikan, 2007: A synthesis of global particle export from the surface ocean and cycling through the ocean interior and on the seafloor. *Global Biogeochemical Cycles*, 21, GB4006, doi:10.1029/2006GB002907

[44] The role of diatoms was highlighted in Bigelow, H.B, 1924, Plankton of the offshore waters of the Gulf of Maine, *Bulletin of the Bureau of Fisheries*, Document 968, 486pp.,

[45] Landry, M.R., M.E. Ondrusek, S.J. Tanner, S.L. Brown, J. Constantinou, R.R. Bidigare, K.H. Coale and S. Fitzwater, 2000: Biological response to iron fertilization in the eastern equatorial Pacific (IronEx-II), I. Microplankton community abundances and biomass, *Mar. Ecol. Prog. Ser.*, 201, 27-42.

[46] Takeda, S. and A. Tsuda, 2005: An in situ iron enrichment experiment in the western subarctic Pacific (SEEDS): Introduction and summary, *Progress in Oceanography*, 64, 95-109.

[47] Coale , K.H., K. S. Johnson, F. P. Chavez, K. O. Buesseler, R. T. Barber, M. A. Brzezinski, W.P. Cochlan, F.J. Millero, P.G. Falkowski, J. E. Bauer, R. H. Wanninkhof, R.M. Kudela, M.A. Altabet, B.E. Hales, T. Takahashi, M. R. Landry, R.R. Bidigare, X. Wang, Z. Chase, P.G. Strutton, G.E. Friederich, M.Y. Gorbunov, V.P. Lance, A.K. Hilting, M.R. Hiscock, M. Demarest, W.T. Hiscock, K.F. Sullivan, S.J. Tanner, R.M. Gordon, C.N. Hunter, V.A. Elrod, S.E. Fitzwater, J.L. Jones, S. Tozzi, M. Koblizek, A.E. Roberts, J. Herndon, J. Brewster, N. Ladizinsky, G. Smith, D. Cooper, D. Timothy, S.L. Brown, K.E. Selph, C.C. Sheridan, B.S. Twining, and Z.I. Johnson., 2004, Southern Ocean iron enrichment experiment: Carbon cycling in high- and low-Si waters, *Science*, 304, 408-414.

[48] Cooper,. D., A.J. Watson, and P.D. Nightingale, 1996: Large decrease in ocean-surface CO2 fugacity in response to in situ iron fertilization, *Nature*, 383, 511-513.

[49] Bakker, D.C.E., A.J. Watson and C.S. Law, Southern Ocean iron enrichment promotes inorganic carbon drawdown, *Deep Sea Res. II*, 44, 91-127.

[50] Tsuda, A., S. Takeda, H. Saito, J. Nishioka, Y. Nojiri, I. Kudo, H. Kiyosawa, A. Shiomoto, K. Imai, T. Ono, A. Shimamoto, D. Tsumune, T. Yoshimura, T. Aono, A. Hinuma, M. Kinugasa, K. Suzuki, Y. Sohrin, Y. Noiri, H. Tani, Y. Deguchi, N. Tsurushima, H. Ogawa, K. Fukami, K. Kuma, and T. Saino. 2003. A mesoscale iron enrichment in the western subarctic Pacific induces a large centric diatom bloom, *Science* 300, 958-961.

[51] Watson, A.J., C.S. Law, K.A. Van Scoy, F.J. Millero, W. Yao, G.E. Friederich, M.I. Liddicoat, R.H. Wanninkhof, R.T. Barber, and K.H. Coale. 1994. Minimal effect of iron fertilization on sea-surface carbon dioxide concentrations, *Nature* 371, 143-145.

[52] Wanninkhof, R. and W.R. McGillis 1999: A cubic relationship between air-sea CO2 exchange and wind speed, *J. Geophys. Res.*, 97, 7383-7382.

[53] De Baar, H.J.W., P.W. Boyd, K.H. Coale, M.R. Landry, A. Tsuda, P. Assmy, D.C.e. Bakker, Y. Bozec, R.T. Barber, M.A. Brzezinski, K.O. Buesseler, M. Boyé, P.L. Croot, F. Gervais, M.Y. Gorbunov, P.J. Harrison, W.T. Hiscock, P. Laan, C. Lancelot, C.S. Law, M. Levasseur, A. Marchetti, F.J. Millero, J. Nishioka, Y. Nojiri, T. van Oijen, U. Riebesell, M.J.A. Rijkenberg, H. Saito, S. Takeda, K.R. Timmermans, M.J.W. Veldhuis, A.M. Waite, C.- S. Wong.2005. Synthesis of iron fertilization experiments: From the Iron Age in the Age of Enlightment, *J. Geophys. Res.*, 110, C09S16, doi:10.1029/2004JC002601.

[54] Hiscock, W.T. and F.J. Millero, Nutrient and carbon parameters during the Southern Ocean iron experiment, SOFeX, *Deep Sea Res. I*, ,52 2086-2108.

[55] Lomas, M. W., D.K. Steinberg, T. Dickey, C.A. Carlson, N.B. Nelson, R.H. Condon and N.R. Bates. Increased ocean carbon export in the Sargasso Sea linked to climate variability is countered by its enhanced mesopelagic attenuation, Biogeosciences, 7, 57-70, doi:10.5194/bg-7-57-2010, 2010.

[56] Lutz, M.J., K. Caldeira, R. B. Dunbar, and M. J. Behrenfeld (2007), Seasonal rhythms of net primary production and particulate organic carbon flux to depth describe the efficiency of biological pump in the global ocean, *J. Geophys. Res.*, 112, C10011, doi:10.1029/2006JC003706.

[57] Trainer, V.L., B.M. Hickey, and S.S. Bates. 2008. Diatoms.. In: Walsh, P.J., S.L. Smith, L.E. Fleming, H. SoloGabriele, W.H. Gerwick (eds.). *Oceans and Human Health: Risks and Remedies from the Sea*, pp.219-238. Elsevier Science Publishers, New York.

[58] Tsuda, A., S. Takeda, H. Saito, J. Nishioka, I. Kudo, Y. Nojiri, K. Suzuki, M. Uematsu, M.L. Wells, D. Tsumune, T. Yoshimura, T. Aono, T. Aramaki, W.P. Cochlan, M. Hayakawa, K. Imai, T. Isada, Y. Iwamoto, W.K. Johnson, S. Kameyama, S. Kato, H. Kiyosawa, Y. Kondo, M. Levasseur, R.J. Machida, I. Nagao. F. Nakagawa, T. Nakanishi, S. Nakatsuka, A. Narita, Y. Noiri, H. Obata, H. Ogawa, K. Oguma, T. Ono, T. Sakuragi, M. Sasakawa, M. Sato, A. Shimamoto, H. Takata, C.G. Trick, Y.W. Watanabe, C.-S. Wong and N. Yoshie. 2007. Evidence for the grazing hypothesis: Grazing reduces phytoplankton responses of the HNLC ecosystem to iron enrichment in the western subarctic Pacific (SEEDS II). *J. Oceanogr.* 63:983-994.

[59] Marchetti A, N.D. Sherry, H. Kiyosawa, A. Tsuda and P.J. Harrison. 2006. Phytoplankton processes during a mesoscale iron enrichment in the NE subarctic Pacific: Part I -biomass and assemblage. *Deep-Sea Res. Part II.* 53:2095-2113.

[60] Chisholm S.W., P.G. Falkowski and J.J. Cullen. 2001. Dis-crediting ocean fertilization. *Science* 294:309-310.

[61] Coale K.H., K.S. Johnson, S.E. Fitzwater, R.M. Gordon, S. Tanner, F.P. Chavez, L. Ferioli, C. Sakamoto, P. Rogers, F. Millero, P. Steinberg, P. Nightingale, D. Cooper, W.P. Cochlan, M. R. Landry, J. Constantinou, G. Rollwagen, A. Trasvina, and R. Kudela. 1996. A massive phytoplankton bloom induced by an ecosystem scale iron fertilization experiment in the equatorial Pacific. *Nature* 383:495-501.

[62] Boyd, P.W., A.J. Watson, C.S. Law, E.R. Abraham, T. Trull, R. Murdoch, D.C.E. Bakker, A.R. Bowie, K.O. Buesseler, H. Chang, M. Charette, P. Croot, K. Downing, R. Frew, M. Gall, M. Hadfield, J. Hall, M. Harvey, G. Jameson, J. LaRoche, M. Liddicoat, R. Ling, M.T. Maldonado, R.M. McKay, S. Nodder, S. Pickmere, R. Pridmore, S. Rintoul, K. Safi, P. Sutton, R. Strzepek, K. Tanneberger, S. Turner, A. Waite, J. Zeldis. 2000. A mesoscale phytoplankton bloom in the polar Southern Ocean stimulated by iron fertilization. *Nature* 407:695-702.

[63] Gervais F, U. Riebesell and M.Y. Gorbunov. 2002. Changes in primary productivity and chlorophyll a in response to iron fertilization in the Southern Polar Frontal Zone. *Limnol. Oceanogr.* 47:1324-1335.

[64] Coale , K.H., K. S. Johnson, F. P. Chavez, K. O. Buesseler, R. T. Barber, M. A. Brzezinski, W.P. Cochlan, F.J. Millero, P.G. Falkowski, J. E. Bauer, R. H. Wanninkhof, R.M. Kudela, M.A. Altabet, B.E. Hales, T. Takahashi, M. R. Landry, R.R. Bidigare, X. Wang, Z. Chase, P.G. Strutton, G.E. Friederich, M.Y. Gorbunov, V.P. Lance, A.K. Hilting, M.R. Hiscock, M. Demarest, W.T. Hiscock, K.F. Sullivan, S.J. Tanner, R.M. Gordon, C.N. Hunter, V.A. Elrod, S.E. Fitzwater, J.L. Jones, S. Tozzi, M. Koblizek, A.E. Roberts, J. Herndon, J. Brewster, N. Ladizinsky, G. Smith, D. Cooper, D. Timothy, S.L. Brown, K.E. Selph, C.C. Sheridan, B.S. Twining, and Z.I. Johnson., 2004, Southern Ocean iron enrichment experiment: Carbon cycling in high- and low-Si waters, *Science*, 304, 408-414.

[65] Peloquin J.A. and W.O. Smith 2006. The role of phytoplankton size on photochemical recovery during the southern ocean iron experiment. *J. Phycol.* 42:1016-1027.

[66] Hoffmann L.J., I. Peeken and K. Lochte 2006. Different reactions of Southern Ocean phytoplankton size classes to iron fertilization. *Limnol. Oceanogr.* 51:1217-1239

[67] Trick, C.G., B.D. Bill, W.P. Cochlan, M.L. Wells, V.L. Trainer, and L.P. Pickell. 2009. Iron enrichment stimulates toxic diatom production in the High Nitrate Low Chlorophyll eastern

Subarctic Pacific. Fifth Symposium on Harmful Algae in the U.S., Ocean Shores, WA, November 2009.

[68] Trick, C.G., B.D. Bill, W.P. Cochlan, M.L. Wells, V.L. Trainer, V. L., and L.P. Pickell. 2009. Understanding complexities in the induction of toxic algal blooms: Domoic acid. GEOHAB: Second Open Science Meeting on Harmful Algal Blooms and Eutrophication, Beijing, China, October 2009.

[69] Trick, C.G., B.D. Bill, W.P. Cochlan, M.L. Wells, V.L. Trainer, and L.P. Pickell. Iron enrichment stimulates toxic diatom production in High Nitrate Low Chlorophyll areas. *Proc. Nat. Acad. Sci.* In press..

[70] Glibert, P.A., J. Harrison, C. Heil and S. Seitzinger, 2006: Escalating worldwide use of urea- A global change contributing to coastal eutrophication, *Biogeochemistry*, 77, 441-463.

[71] Bryan, K. and M.D. Cox, 1967. A numerical investigation of the oceanic general circulation, *Tellus*, 19, 54-80.

[72] Matusmoto, K., J.L. Sarmiento, R.M. Key, O. Aumont, J.L. Bullister, K. Caldeira, J.M. Campin, S.C. Doney, H. Drange, J.C. Dutay, M. Follows, Y. Gao, A. Gnanadesikan, N. Gruber, A. Ishida, F. Joos, K. Lindsay, E. Maier-Reimer, J.C. Marshall, R.J. Matear, P. Monfray, R. Najjar, G.K. Plattner, R, Schlitzer, R. Slater, P.S. Swathi, I.J. Totterdell, M.F. Weirig, Y. Yamanaka, A. Yool and J.C. Orr. 2004. Evaluation of ocean carbon cycle models with data-based metrics, *Geophys. Res. Lett.*, 31, L07303, doi:10.1029/2003GL018970.

[73] Najjar, R.G. 1990. Simulations of the phosphorus and oxygen cycles in the world ocean using a general circulation model, Ph.D. Thesis, Princeton Univ.

[74] Sarmiento, J.L., R.D. Slater, M.J. Fasham, H.W. Ducklow, J.R. Toggweiler, and G.T. Evans. 1993. A seasonal three-dimensional ecosystem model of nitrogen cycling in the North Atlantic euphotic zone. *Global Biogeochemical Cycles*, 7(2), 417-450.

[75] Examples include Moore, J.K., S.C. Doney and K. Lindsay. 2004. Upper ocean ecosystem dynamics and iron cycling in a global 3-D model, *Global Biogeochemical Cycles*, 18, GB4028, doi:10.1029/2004GB002220. Aumont, O. and L. Bopp. 2006. Globalizing results from in-situ iron fertilization studies, *Global Biogeochemical Cycles.*, 20, GB2017, doi:10.1029/2005GB002591, Galbraith, E.D., A. Gnanadesikan, J.P. Dunne and M.R. Hiscock. 2009. Regional impacts of iron-light collimitation in a global biogeochemical model, *Biogeosciences Discussions*, 5, 7517- 7564.

[76] Sarmiento, J.L. and J.C. Orr. 1991. Three-dimensional simulations of the impact of Southern Ocean nutrient depletion on atmospheric CO2 and ocean chemistry, *Limnol. Oceanogr.* 36, 1928-1950.

[77] Cao, L., and K. Caldeira. 2010. Can ocean iron fertilization mitigate ocean acidification? *Climatic Change*, doi:10.1007/s10584-010-9799-4.

[78] Marinov, I., A. Gnanadesikan, J.R. Toggweiler and J.L Sarmiento. 2006. The Southern Ocean biogeochemical divide, *Nature*, 441, doi:10.1038/nature04883/

[79] Gnanadesikan, A. and I. Marinov. 2008. Export is not enough: Nutrient cycling and carbon sequestration, *Mar. Ecol. Prog. Series,* 364, 289-294.

[80] Parekh, P., M.J. Follows, S. Dutkiewicz and T. Ito. 2006. Physical and biological regulation of the soft tissue carbon pump, *Paleoceanography*, 21,PA3001, doi:10.1029/2005PA001258.

[81] Aumont, O. and L. Bopp. 2006. Globalizing results from in-situ iron fertilization studies, *Global Biogeochemical Cycles.*, 20, GB2017, doi:10.1029/2005GB002591

[82] Sarmiento, J.L., R.D. Slater, J.P. Dunne, A. Gnanadesikan and M.R. Hiscock. 2009. Efficiency of small-scale carbon mitigation by patch iron fertilization, *Biogeosciences Discussions*, 6, 10381-10446.

[83] Archer, D.A. 2005. Fate of fossil fuel CO2 in geologic time. *J. Geophys. Res*. VOL. 110, C09S05, doi:10.1029/2004JC002625, 2005

[84] Gnanadesikan, A., J.L. Sarmiento and R.D. Slater. 2003. Effects of patchy ocean fertilization on atmospheric carbon dioxide and biological production, *Global Biogeochemical Cycles*, 17, 1050, doi:10.1029/2002GB001940. See also Sarmiento, J.L., R.D. Slater, J.P. Dunne, A. Gnanadesikan and M.R. Hiscock. 2009. Efficiency of small-scale carbon mitigation by patch iron fertilization, *Biogeosciences Discussions*, 6, 10381-10446.

[85] Arrigo K.R., R.B. Dunbar, M. Lizotte and D.H. Robinson. 2002.Taxon-specific differences in C/P and N/P drawdown for phytoplankton in the Ross Sea, Antarctica. *Geophy. Res. Lett.* 29, 1938, doi:10.1029/2002GL015277.

[86] Gnanadesikan, A. and I. Marinov. 2008. Export is not enough: Nutrient cycling and carbon sequestration, *Mar. Ecol. Prog. Series,* 364, 289-294.

[87] Gnanadesikan, A., J.L. Sarmiento and R.D. Slater. 2003. Effects of patchy ocean fertilization on atmospheric carbon dioxide and biological production, *Global Biogeochemical Cycles*, 17, 1050, doi:10.1029/2002GB001940

[88] Marinov, I., A. Gnanadesikan, J.R. Toggweiler and J.L Sarmiento. 2006. The Southern Ocean biogeochemical divide, *Nature*, 441, doi:10.1038/nature04883

[89] Aumont, O. and L. Bopp. 2006. Globalizing results from in-situ iron fertilization studies, *Global Biogeochemical Cycles.*, 20, GB2017, doi:10.1029/2005GB002591

[90] Jin, X., and N. Gruber. 2003. Offsetting the radiative benefit of ocean iron fertilization by enhancing N2O emissions, *Geophysical Research Letters*, 30(24), 2249, doi:10.1029/2003GL018458

[91] This is one of the fundamenatal conclusions from: Lutz, M.J., K. Caldeira, R.B. Dunbar, and M.J. Behrenfeld. 2007. Seasonal rhythms of net primary production and particulate organic carbon flux describe biological pump efficiency in the global ocean. *J. Geophys. Res., 112*, C10011, doi:10.1029/2006JC003706

[92] Cao, L., and K. Caldeira. 2010. Can ocean iron fertilization mitigate ocean acidification? *Climatic Change*, doi:10.1007/s10584-010-9799-4.

[93] International Maritime Organization, Report Of The 30th Meeting of the Scientific Group of the London Convention and the 1st Meeting of the Scientific Group of the London Protocol, LC/SG 30/14, Sections 2.22-2.28, 25 July 2007

[94] International Maritime Organization, Report Of The 31st Meeting of the Scientific Group of the London Convention and the 2nd Meeting of the Scientific Group of the London Protocol, LC/SG 31/16, Annex 2, 7 July 2008.

[95] International Maritime Organization, Report Of The 30th Consultative Meeting and the 3rd Meeting of Contracting Parties, LC 30/16, Annex 6, 9 December 2008

In: Climate, Energy, and Environment. Vol. 1 ISBN: 978-1-62948-407-5
Editor: Jake S. Diaz

Chapter 3

THE LACEY ACT: PROTECTING THE ENVIRONMENT BY RESTRICTING TRADE[*]

Kristina Alexander

SUMMARY

The Lacey Act was enacted in 1900 to prevent hunters from illegally killing game in one state and escaping prosecution by crossing state lines. It has evolved into a law that prohibits import, export, transport, purchase, or sale of species when that action would violate state, federal, tribal, or foreign law. Congress amended the Lacey Act most recently in 2008, expanding the reach of the act to include timber and timber products. Implementation of the 2008 Amendments has proved controversial, and the Department of Agriculture Animal and Plant Health Inspection Service (APHIS) initially delayed implementing the act's new declaration requirements for importing wood products.

Some find the Lacey Act puzzling. While people charged with violating the act are charged with violating a U.S. law, that prosecution is premised on a violation of another law, sometimes the law of another country. That has led some to claim that the United States is enforcing the laws of another country. U.S. conservation laws (such as the Lacey Act), however, have long protected species and habitats even outside of the United States. Worldwide conservation was one reason for expanding Lacey Act coverage to more plants in 2008. Preserving U.S. timber jobs

[*] This is an edited, reformatted and augmented version of Congressional Research Service, Publication No. R42067, dated July 16, 2013.

and prices was another reason. However, the 2008 Amendments to the Lacey Act allow enforcement of foreign laws that are not directly related to conservation or U.S. jobs, such as failure to pay foreign stumpage fees, or shipping wood in violation of a country's export restrictions. After search warrants were executed by the Department of the Interior Fish and Wildlife Service (FWS) against Gibson Guitar Corp. of Nashville, TN, in August 2011, based on the possible illegal import of wood from India, the 112th Congress took another look at whether the 2008 Amendments achieve the goals of the Lacey Act.

In July 2012, Gibson settled charges that imports in 2008 and 2009 had violated the Lacey Act, agreeing to pay a $300,000 penalty, make a $50,000 donation to a wildlife nonprofit group, and withdraw its claims to recover wood forfeited during the Madagascar investigation. The United States returned the wood from India seized in 2011. As part of the settlement, the United States agreed that Indian trade law, on which the alleged 2011 violations were based, was unclear. However, Gibson admitted that it had continued to import ebony from Madagascar after learning that it was illegal.

BACKGROUND OF THE LACEY ACT

The Lacey Act addresses conservation by targeting trade. As adopted in 1900, the Lacey Act made it a federal crime to ship game killed in violation of one state's laws to another state.[1] It also regulated the introduction of non-native species and prohibited shipment of wildlife without clear identification of the contents and shipper.[2] This report will examine the portions of the Lacey Act that pertain to transporting fish, wildlife, and plants that were taken in violation of another law, and will not address the other provisions of the act.

While it has been amended many times since 1900, the Lacey Act's format has remained the same: with respect to fish, wildlife, and plants, generally speaking, it is a violation of the act to violate another law, be it state, federal, tribal, or foreign. Accordingly, making a case under the Lacey Act requires a violation of an underlying law as well as a violation of the act itself.

The act has a broad reach. It applies to any person, with *person* defined to include individuals, corporations, government officials, and government agencies.[3] It covers violations of state, federal, tribal, or foreign laws (including regulations) that regulate "the taking, possession, importation, exportation, transportation, or sale" of wildlife or plants.[4] It is enforced by the Fish and Wildlife Service (FWS) of the Department of the Interior, the National Oceanic and Atmospheric Administration of the Department of

Commerce, Customs and Border Protection (CBP) of the Department of Homeland Security, and the Animal and Plant Health Inspection Service (APHIS) and the Forest Service of the Department of Agriculture.[5]

The 2008 Amendments

In 2008, the Lacey Act was amended, expanding the definition of plants to include trees and plant products, and prohibiting different types of trade related to plants. In August 2011, FWS executed search warrants for the premises of Gibson Guitar Corp. reportedly based on Gibson's violations of the 2008 Amendments when Gibson allegedly imported unfinished wood in violation of India's export laws (laws the U.S. government later conceded were vague).[6] Because Gibson has claimed its imported wood was sustainably harvested, some argue that the 2008 Amendments do not conserve environmental resources exclusively but can hurt U.S. workers by enforcing foreign laws that protect foreign workers. Others argue that the act protects U.S. jobs.

Legislative History of the 2008 Amendments

According to the legislative history for the 2008 amendments, its purposes were two-fold: to promote ecosystem protection and to aid the U.S. timber industry. According to the House conference report, the amendments address the following problems:

> [I]llegal logging undermines responsible forest enterprises by distorting timber markets with unfair competition and price undercutting. Illegal logging also threatens the conservation of forest resources, wildlife, and biodiversity, by facilitating forest conversion to non-forest uses and depleting or completely eliminating certain forest ecosystems or the habitat of certain forest dependent wildlife. Finally, illegal logging results in a loss of revenue when taxes or royalties are not paid that could otherwise be invested in sustainable forest management or economic development.[7]

Senator Wyden, a co-sponsor of a bill that led to the 2008 Amendments, described the benefits of the act as follows: "This legislation will raise the risks for illegal trade without harming legal trade and will be an important step toward leveling a playing field currently stacked against the U.S. forest products industry and importers and retailers committed to trading in legal wood products."[8]

Senator Alexander, another co-sponsor, discussed the bill's projected economic impact of illegal timber imports: "[I]t is an estimated $1 billion a year in depressed prices and reduced exports. It depresses prices $500 million to $700 million annually. It means the people who play by the rules in the United States are having money taken from them by criminals who don't play by the rules in other countries ..."[9]

LACEY ACT PROHIBITIONS RELATED TO WILDLIFE

Because of the complexity of the act, involving federal, state, and/or foreign laws, the act is best analyzed with respect to the type of goods it seeks to protect. This report will first look at those parts of the Lacey Act pertaining to wildlife. Plants will be discussed later in the context of the 2008 Amendments.

However, as will be discussed in greater detail below, with respect to both wildlife and plants, penalties for Lacey Act violations are based on the degree of knowledge or care taken by the alleged offender, while forfeiture of protected wildlife and plants under the Lacey Act is on a strict liability basis.

Violations of U.S. or Tribal Law

The act uses the term *fish or wildlife* to include "any wild animal, whether living or dead ... [including] any part, product, egg, or offspring thereof."[10] This report will simply use the term *wildlife*. There are two types of prohibited acts pertaining to wildlife.

The first makes it illegal to "import, export, transport, sell, receive, acquire, or purchase" wildlife if that creature was "taken, possessed, transported, or sold in violation of any law ... of the United States or in violation of any Indian tribal law."[11] In this case, a violation of a federal law that regulates wildlife is also a violation of the Lacey Act if the wildlife in question was imported, exported, transported, sold, or received.

Example[12] 1 (Section 3372(a)(1))

Assume a person has killed a hummingbird in the state of Oklahoma. That is a violation of the Migratory Bird Treaty Act (MBTA), a federal law.[13] However, it is not a violation of the Lacey Act. Even though wildlife was taken in violation of a federal law, that wildlife was not then imported, exported, transported, sold, received, acquired, or purchased, as required for a Lacey Act violation.

Example 2 (Section 3372(a)(1))

Assume a person has killed a hummingbird in the state of Oklahoma. That person has violated the MBTA. That person then sells the feathers to a person in Oklahoma. By making the sale, that person has also violated the Lacey Act because there was a sale of wildlife (which by definition includes parts of that animal) that was taken in violation of a federal law. The buyer also has violated the Lacey Act because the buyer purchased wildlife that was taken in violation of a federal law.

Violations of State or Foreign Law

The second type of wildlife prohibition under the Lacey Act involves interstate or foreign commerce. Section 3372(a)(2)(A) of the Lacey Act makes it illegal to "import, export, transport, sell, receive, acquire, or purchase in interstate or foreign commerce" wildlife if that wildlife was "taken, possessed, transported, or sold in violation of any law or regulation of any State or in violation of any foreign law."[14] For this type of violation, the underlying law must be state or foreign. While the language clearly includes state regulatory violations as a predicate act to a Lacey Act violation ("in violation of any law or regulation of any State"), it is not so clear from the statute whether violations of foreign regulations may also be the basis for a Lacey Act violation. Courts, however, have held that foreign regulations may be the foundation for a charge. The Eleventh Circuit reviewed congressional intent before holding that violations of foreign regulations and other legally binding provisions are also a basis for making a Lacey Act enforcement case.[15] The Ninth Circuit found that Congress used the term "foreign law" so it would encompass the many forms that foreign legal edicts may take.[16]

Example 3 (Section 3372(a)(2)(A))

Assume a person has killed a deer in violation of Wisconsin's hunting laws. The deer is field dressed and consumed without leaving the state or entering commerce. There is no violation of the Lacey Act because there is no interstate commerce.

Example 4 (Section 3372(a)(2)(A))

Assume a person has killed a deer in violation of Wisconsin's hunting laws by hunting in an area where it is not allowed—for example, a petting zoo. The venison is sold to a person in interstate commerce. Because interstate commerce was involved, the seller has violated the Lacey Act. Likewise, the buyer also has violated the Lacey Act.

The Lacey Act has an expanded use of the term *sale* in the context of wildlife violations. It deems an offer to provide guiding or outfitting services, or to provide a hunting or fishing license or permit as a sale.[17] Accordingly, obtaining a hunting license in violation of tribal law can be the predicate violation of the Lacey Act.[18]

Before a violation occurs under either Section 3372(a)(1) or Section 3372(a)(2)(A), both parts of the Lacey Act must be triggered—the underlying or predicate law and then the Lacey Act. Courts must determine the validity of both the predicate violation and the action that triggers the Lacey Act charge. Where the predicate violation is one of foreign law or regulation, the federal courts have the authority to determine whether a foreign law has been violated.[19] Federal courts also have jurisdiction when the underlying violation is one of tribal law.[20]

Establishing violations of foreign law can lead to complicated situations. For example, the United States seized the hide and horns of an Afghan Urial, a type of big horn sheep, imported to Virginia from Pakistan. The importer of the animal argued that Pakistan had no national law protecting the Urial and that he had a valid permit from the province where it was taken. The Fifth Circuit held that a Pakistani law that prohibited the export of wild animal hides barred the export of the Urial pelt.[21] In another case a man was convicted of importing spiny lobsters to the United States that were caught in violation of Honduras law. By the time the conviction was appealed, "the Honduran government reversed its position; it [refuted] the validity of the laws it previously verified."[22] The Eleventh Circuit held that Honduras law at the time the conduct occurred prohibited harvesting the spiny lobsters and upheld the Lacey Act conviction.[23]

In determining that the predicate violation occurred, the court does not have to find that the procedures of that jurisdiction were followed. Federal procedural rules apply.[24] For example, the Fourth Circuit has held that the general federal statute of limitations applies.[25] For a Lacey Act violation, there is a five-year statute of limitations.[26]

LACEY ACT PROHIBITIONS RELATED TO PLANTS

The Lacey Act definition of plant was expanded in 2008 to include any wild member of the plant kingdom including roots, seeds, parts, or products, including trees.[27] Prior to the 2008 Amendments, the Lacey Act applied only to those plants that were indigenous to the United States and listed under one of the following: the Endangered Species Act (ESA); the Convention on International Trade in Endangered Species of Wild Fauna and Flora (CITES); or a state law to conserve species threatened with extinction.[28] It did not expressly apply to plant products. The revised definition has some exclusions, such as for scientific research and for plants that will be transplanted.[29] The Lacey Act prohibits four types of acts related to plants and requires submission of a declaration for importing plants.

Violations of U.S. or Tribal Law

One prohibited activity is identical for plant violations as it is for wildlife violations. The Lacey Act makes it illegal to "import, export, transport, sell, receive, acquire, or purchase" plants if the plant was "taken, possessed, transported, or sold in violation of [U.S. or Indian tribal law]."[30] As for the wildlife violations, these cases tend to be fact-specific. For example, the saguaro cactus, an icon of the American Southwest, is not protected under federal law but is under Arizona state law. Accordingly, taking a saguaro cactus would not be a Lacey Act violation under Section 3372(a)(1),[31] but would be under Section 3372(a)(2)(B), as discussed below.

Example 5 (Section 3372(a)(1))

A person in New York found sandplain gerardia on federal lands, a plant listed as endangered under the ESA. That person kills the plant. The person has violated the ESA,[32] but has not violated the Lacey Act because there has been no act of import, export, transport, sale, receipt, acquisition, or purchase of the plant.

Example 6 (Section 3372(a)(1))

A person digs up a Kuenzler hedgehog cactus (an endangered species) from a state park, which violates state law, and sells it to someone in another state. The buyer plants it in the perfect spot and it flourishes. Even though the plant does not die, there is still a violation of the ESA, which prohibits digging up endangered plants in violation of state law, as well as buying or selling them in interstate commerce.[33] Therefore, both the seller and the buyer have violated the ESA. Because there was a sale, both have also violated the Lacey Act.

Violations of State or Foreign Law Protecting Plants

The 2008 Amendments added three other prohibited acts related to plants. The first directly relates to laws that conserve plants. It prohibits importing, exporting, transporting, selling, receiving, acquiring, or purchasing any plant in interstate or foreign commerce if that plant was taken, possessed, transported, or sold in violation of state or foreign laws that

1. protect plants;
2. regulate the theft of plants;
3. regulate the taking of plants from a park, forest reserve, or other officially protected area;
4. regulate the taking of plants from an officially designated area; or
5. regulate the taking of plants without, or contrary to, required authorization.[34]

Under this section, Section 3372(a)(2)(B)(i), it is illegal to take a plant, including a tree or a plant product, in interstate commerce, if the state or foreign country protected the specific plant in some way or protected the area from which the plant was taken, and that protection was violated. Under the 2008 Amendments, taking a saguaro cactus, as discussed above, would be prosecuted under Section 3372(a)(2)(B).

Example 7 (Section 3372(a)(2)(B)(i))

A person takes wild sage leaves from the roadside of Turkey, and then brings the leaves into the United States for the person's own use. Assume it is against Turkish law to pick sage leaves off roadside plants without authorization. A Lacey Act violation occurred because the person imported sage leaves that violated a foreign law.

Example 8 (Section 3372(a)(2)(B)(i))

A person picks leaves of a plant that grows only in Turkey from the roadside and then brings the leaves into the United States to sell. Assume this violates Turkish law. Assume it is against U.S. law to sell those leaves. Prior to enactment of the 2008 Amendments there would be no Lacey Act violation, as the definition of *plant* at the time required that the plant be indigenous to the United States (or listed under ESA, CITES, or state law) for the prohibitions to apply, and underlying foreign law was not applicable in the pre-2008 Lacey Act.

Violations of State or Foreign Law Requiring Fees

The other two plant prohibitions are less directly linked to conservation. One prohibition under the 2008 Amendments makes it illegal to import, export, transport, sell, receive, acquire, or purchase a plant in interstate or foreign commerce that was "taken, possessed, transported, or sold without the payment of appropriate royalties, taxes, or stumpage fees required for the plant by any law or regulation of any State or any foreign law."[35] Under this provision, the type of plant or the location of the plant is not at issue. The predicate act is based on whether the plant was harvested legally in relation to required fees. Prohibiting import, export, sales, and purchases of timber based on payment of fees could be argued as having a conservation interest by limiting illegally harvested trees, even if the illegality is related to paying fees, rather than the type or locale of the trees. The theory that this prohibition supports conservation is that lawless harvesting limits the economic advantage timber harvesting typically brings, such as funds for reforestation or other protections.[36] From an enforcement perspective, these violations can be easier to prove because there will be paperwork showing whether the fees were paid (as opposed to proof of illegal origin, for example), and the conservation benefits can more easily be realized.

Example 9 (Section 3372(a)(2)(B)(ii))

An American furniture manufacturer imports oak from Canada knowing that the price is good because the timber company selling the wood "knows a guy" who can avoid paying fees for cutting the trees. The American furniture company has violated the Lacey Act even though it was the timber company or "the guy" that did not pay the fees. The act prohibits purchase of timber taken or sold without paying appropriate taxes and royalties. It does not have to be the purchaser or importer that was required to pay the fees.

Violations of State or Foreign Law Restricting Export

The last prohibition added by the 2008 Amendments makes it illegal to import, export, transport, sell, receive, acquire, or purchase a plant in interstate or foreign commerce that was "taken, possessed, transported, or sold in violation of any limitation under any law or regulation of any State, or under any foreign law, governing the export or transshipment of plants."[37] Because there is no specific requirement that the export law relate to conservation, some could contend this provision is more likely to preserve trade advantages, rather than to protect ecosystems. But the argument could be made that any restriction of trade in wood and wood products provides conservation benefit by potentially limiting the amount of wood that is harvested and exported. These export laws can serve as an enforcement "choke point" for that jurisdiction. Local officials can moderate how the wood being processed was taken, thus serving a conservation purpose. For example, by reviewing wood finishing, the foreign government has the opportunity to review harvesting issues, such as location and types of wood. Another conservation benefit to enforcing foreign export laws can be illustrated by the example of a country with a limited supply of a type of wood but an unlimited demand for that wood worldwide. If processing were allowed to occur outside of the country, that rare timber would be harvested without any limitation. By requiring local processing, harvest is limited to the speed of domestic manufacturing, thus conserving the timber. These provisions may also aid U.S. timber industry interests by restricting imported wood and wood products.

Example 10 (Section 3372(a)(2)(B)(iii))

Under the law of a foreign country, only hand-made toothpicks may be exported. That same country has no restrictions on harvesting a common tree. A U.S. importer imports machine-made toothpicks, made of that common tree. The importer has violated the Lacey Act because it has imported plants (or plant products) in violation of a foreign law governing export of plants.

Declaration

The 2008 Amendments also added a requirement for a declaration to be prepared by those importing plants under the act. Under Section 3372(f), after a period of time following enactment of the amendments (which occurred May 22, 2008), it became illegal to import a plant (or plant product) without a declaration showing the following:

- the scientific name (including genus and species);
- the value;
- the quantity, including unit of measure; and
- the country from which the plant was taken.[38]

The declaration is submitted by the importer at the time of import. A copy of the form is maintained in either electronic or paper form by APHIS and CBP.

The 2008 Amendments required the declaration as of December 15, 2008. However, in response to concerns about the complexity of implementing the declaration requirement, especially for importers of wood products, APHIS established a delayed enforcement schedule, setting compliance stages for some more complicated products for April 1, 2010.[39] Certain products, such as wooden furniture, still are not on an enforcement schedule. (See the **Appendix** for the Schedule.) While the declarations were required for all products as of December 15, 2008, APHIS said it would not prosecute for failure to have a declaration until the dates indicated in the implementation schedule.[40]

The 2008 Amendments also required APHIS to evaluate the declaration process within two years of enactment.[41] In February 2011, APHIS published a notice that it was initiating its review and seeking comments on the implementation of declarations.[42] A report must be submitted to Congress within six months of completing the review.[43] The report was submitted to Congress on May 21, 2013.[44]

Lacey Act Penalties: Civil Violations, Felonies, Misdemeanors, and Forfeiture

While Section 3372 states what actions are not legal, Section 3373 tells what actions may be punished. The Lacey Act provides for civil and criminal penalties as well as forfeiture of the protected item. Some conduct can be subject to either civil or criminal charges. The choice is left to the prosecutors at the Department of Justice. These provisions took shape when the Lacey Act was significantly amended in 1981.[45]

Civil Violations

For the violations described in the examples above (as opposed to marking or labeling violations) the standard for a civil violation is whether the person engaged in that conduct "in the exercise of due care should know that the fish or wildlife or plants were taken, possessed, transported, or sold in violation of, or in a manner unlawful under, any underlying law, treaty, or regulation."[46] If so, that person may be assessed "a civil penalty by the Secretary of not more than $10,000 for each such violation."[47] The penalty is reduced if the value of the wildlife or plants is less than $350.[48]

The *exercise of due care* is a common legal phrase meaning the amount of attention a reasonable person in the same circumstances would use. Under this standard, a first-time buyer of imported animals, for example, is likely to be found to have less responsibility than an importer of those animals. It does not excuse deliberate ignorance. Due care requires what is reasonable, such as asking questions. The Senate report supporting the 1981 Amendments described the standard as follows: "Due care simply requires that a person facing a particular set of circumstances undertake[s] certain steps which a reasonable man would take to do his best to insure that he is not violating the law."[49]

According to that Senate report, the due care requirement was added to avoid overzealous prosecution of unknowing violators: "The civil penalty provisions are included with the understanding that they will not be administered to penalize innocent purchasers or consumers, but rather with the clear intent that they will be applied fairly in an equitable and nonabusive manner."[50]

Misdemeanors

Similarly, the Lacey Act makes it a criminal misdemeanor for someone "who knowingly engages in conduct prohibited by any provision of this chapter,"[51] and "in the exercise of due care should know that the fish or wildlife or plants were taken, possessed, transported, or sold in violation of, or in a manner unlawful under, any underlying law, treaty or regulation."[52] The criminal misdemeanor provision explicitly requires that the action that triggers the Lacey Act violation be *knowingly* done. This means that the person must know he (or she) is doing the action, such as transporting the item or buying a wildlife or plant product, not that he knows he is doing something wrong.

For the predicate act, the standard for a misdemeanor, like the civil provision, is based on a *due care* standard: it applies to those who "in the exercise of due care should know" that something illegal was occurring. According to that same 1981 Senate report, this provision was included to avoid leaving "too much potential for abuse and indiscriminate enforcement efforts."[53]

Returning to *Example Two*, above, for illustration, proving the Lacey Act portion of the illegal feather purchase would require the buyer to know only that he was buying a wildlife or plant product, not that he knew he was buying illegal hummingbird feathers. But the predicate violation must also be proved—a court must find that the buyer in the exercise of due care should know that the product was somehow illegal.

Example 2A—Expanded to Consider Misdemeanor Culpability

A person buys a bag of feathers that includes some hummingbird feathers mixed in with legal goose down. It was the same price as every other bag of down. The person knowingly bought feathers, but in the exercise of due care he would have no knowledge that the feathers he bought were illegal.

It is unlikely that a court would find the person by exercising due care should have known he was buying feathers that were taken in violation of federal law.

Example 2B—Expanded to Consider Misdemeanor Culpability

A person buys three tiny green feathers for the same price as a whole bag of down. This person also knowingly committed the act that triggers the Lacey Act by buying the feathers. It is likely that a court would find that this person when buying those tiny feathers did not exercise due care to discover whether a law had been violated. Two factors a court might expect a reasonable person to consider under the circumstances are that the feathers were unusual and that they were more expensive. The failure to ask reasonable questions could support a finding that the second buyer engaged in conduct prohibited by the act (buying feathers) and that any due care would have revealed that the feathers were taken in violation of law, establishing a Lacey Act violation.

A misdemeanor may be punished by up to one year in jail, in addition to or instead of the fine. Despite the Lacey Act authorizing a maximum $10,000 fine for the misdemeanor, under the Criminal Fines Improvement Act, the maximum criminal penalty is adjusted to $100,000 for an individual, and $200,000 for an organization.[54]

Felonies

The felony provision has a higher standard of culpability. Like the misdemeanor provision, it also requires the actor to know he or she was committing the action that triggers the Lacey Act violation—importing, exporting, engaging in the conduct to sell or purchase or attempt to sell or purchase an item taken in violation of another law—but it also requires that the actor know "that the fish or wildlife or plants were taken, possessed, transported, or sold in violation of, or in a manner unlawful under, any underlying law, treaty or regulation."[55] To be a felony there must be knowledge and one of two things: import or export;[56] or commercial conduct and a value of over $350.[57] These requirements were included to avoid imposing big penalties on smaller offenders, according to the 1981 Senate report.[58] Additionally, that report indicates congressional intent to allow innocent violators to escape criminal prosecution, "the act's criminal culpability requirement assures that innocent violators of the act will not be subject to criminal penalties."[59] Punishment for committing a felony includes incarceration of no more than five years, a fine of $250,000 for individuals and $500,000 for organizations, or both.[60]

To illustrate the felony provisions, consider *Example 4*, above, in which an Illinois buyer purchased a deer killed in violation of Wisconsin law. The court must find the buyer knew two elements: (1) that he was buying something; and (2) that the venison he was buying was illegal in some way.

Example 4A—Expanded to Consider Felony Culpability

The buyer purchased the venison at a farmers' market in November (a time when most, if not all states, allow deer hunting). He paid the going rate for the meat. Under this example, a court likely would find that the buyer had not criminally violated the Lacey Act. Everything was consistent with a bona fide purchase. A reasonable person would not know in these circumstances that the venison was tainted, nor in the exercise of due care would discover the underlying violation of law.

Example 4B—Expanded to Consider Felony Liability

The buyer, an avid deer hunter, purchased the fresh venison at midnight in the parking lot of a petting zoo where remains of the deer could be seen in a pile. The price was very cheap and the seller was wearing a mask. It seems likely a court would find the buyer had criminally violated the Lacey Act. The circumstances of the sale/purchase would cause a reasonable person to know something was wrong.

Forfeiture

The Lacey Act also authorizes forfeiture of the illegal items. Forfeiture is where the government confiscates an item involved in a crime.[61] As the Ninth Circuit found in *United States v. 144,744 Pounds of Blue King Crab*, under the law, the illegally taken or imported item is considered contraband, even if possessing it would not otherwise be illegal.[62] If the forfeiture action is civil, the court does not have to find a person liable for wrongdoing. For example, in the case discussed earlier of the Urial (the big horn sheep from Pakistan), the suit was not against the importer for violating the Lacey Act, but was a case where the government seized the hide and horns from the importer.[63] The prosecution was against the contraband itself. Cases in which civil forfeiture is the only issue have case names where the item is named as the defendant. For criminal forfeiture actions, a successful suit requires conviction of the person who owned the contraband.

Forfeiture can serve two purposes: remedial or punitive. According to the Supreme Court, the remedial purposes include preventing forbidden merchandise from circulating, and reimbursing the government for investigation and enforcement expenses.[64] The punitive goals of forfeiture apply to criminal proceedings.

The Lacey Act applies civil forfeiture proceedings to "all fish or wildlife or plants imported, exported, transported, sold, received, acquired, or purchased contrary to [Section 3372 of the act] ... notwithstanding any culpability requirements for civil penalty assessment or criminal prosecution [under Section 3373 of the act]."[65] In situations where the wildlife or plants were part of a "criminal violation of this Act for which a felony conviction is obtained," the act allows civil forfeiture of the "vessels, vehicles, aircraft, and other equipment" used in committing the crime.[66]

Prior to the 2008 Amendments, courts held that there was no innocent owner defense to forfeiture under the Lacey Act. These decisions are based on the act's language that that "*all* fish or wildlife or plants [that violate Section 3372] shall be subject to forfeiture ..."[67] According to the 1981 Senate report, forfeiture was to occur regardless of culpability: "The strict liability forfeiture section of these amendments would allow the protection of various species from harmful illegal trade by withdrawing illegal shipments from the marketplace even when the violation itself is inadvertent."[68]

The only defense described in that report was where the merchandise was without taint and the import ran afoul of the law only in a "minor, technical" manner.[69]

The Fifth Circuit pointed to that legislative background in finding that "the legislative history establishes that the forfeiture statute provides for strict liability, thereby eliminating any 'innocent owner' defense."[70] A federal court in Florida held that the Lacey Act does not provide for an innocent owner defense.[71] As referenced above, in *United States v. 144,744 Pounds of Blue King Crab*, the Ninth Circuit has held that Lacey Act goods are contraband.[72]

The 2008 Amendments added the statement that civil forfeiture was to be "in accordance with the procedure established under [18 U.S.C. Sections 981-986]."[73] Under this statute, the Civil Asset Forfeiture Reform Act (CAFRA),[74] the government has the burden of proof by a preponderance of evidence.[75] The person owning the forfeited goods may also seek immediate release of the property pending trial.[76]

CAFRA allows a defense to a civil forfeiture for innocent owners, bona fide purchasers, and heirs.[77] However, under CAFRA, the innocent owner defense cannot be used for contraband. The exclusion states: "no person may assert an ownership interest under this subsection in contraband or other property that it is illegal to possess."[78] As discussed above, the Ninth Circuit found that goods taken in violation of the Lacey Act are contraband. Additionally, the 1981 Senate report described the goods in the same way: "The act provides for forfeiture of the fish, wildlife and plants on a strict liability basis because the merchandise is, in effect, contraband."[79]

A statement in the House report on the 2008 Amendments further references the *Blue King Crab* decision to note that incorporating CAFRA by reference may not provide a defense to the strict liability forfeiture provisions under the Lacey Act because of the ruling that goods are contraband under the act.[80]

However, it is possible to have the goods returned while the matter is still administrative, before it is referred to court. This is known as remission. Department of the Interior regulations that apply to forfeitures allow the possibility of remission in administrative forfeitures in rare circumstances. When FWS notifies a person that goods are forfeited, that person may file a petition for remission rather than a claim, which would refer the matter to court. Under 50 C.F.R. Section 12.24(f), the regulations provide that "If the Solicitor finds the existence of such mitigation circumstances," goods may be remitted. The only published decision available regarding forfeiture of plant products following the 2008 Amendments considered such a petition before rejecting it.[81]

According to that administrative case, the Solicitor considered "whether the facts demonstrate a person's honest and good faith intent to comply with the law, their diligence and efforts to comply, and whether the facts tend to show a lack of negligence or moral culpability for whatever failure or mistake is involved in the non-compliance ..."[82] This administrative forfeiture process appears consistent with the legislative discussion of allowing release of untainted goods related to innocent trade violations.

Example 11—Forfeiture of Contraband

While visiting Nicaragua, a couple buys a table made of a rare wood protected under U.S. law. When they enter the United States they lie on the customs declaration, saying the table was a different, non-protected species. They have violated the Lacey Act under Section 3372(a)(1), by transporting an item in violation of U.S. law by lying on a customs declaration. In addition to any civil or criminal prosecution against the couple, the table is contraband and may be seized by the government.

GIBSON GUITAR AND THE 2008 AMENDMENTS

In August 2011, FWS executed search warrants at Gibson Guitar Corp. premises and seized wood imported from India. Reportedly, the FWS agents were armed. However, no charges were brought against Gibson or its officers. According to a statement from Gibson, the wood was certified by the Forest Stewardship Council[83] as sustainably harvested and was seized by FWS because it was "not finished by Indian workers."[84] FWS and the Department of Justice (DOJ) did not comment on the case, although the government settled this and other charges against Gibson, with Gibson paying a fine, and the government agreeing that India laws regarding exporting ebony and rosewood fingerboard blanks were vague.[85] The wood was returned to Gibson.

Section 3375(b) of the Lacey Act authorizes federal agents to carry firearms when enforcing the act. It also allows execution of search and arrest warrants, including searching any vessels, packages, containers, documents, or permits related to the action.

Background

To make sense of the statements in the affidavit prepared to obtain the search warrants, it helps to have some background of relevant trade terminology. Countries generally use the International Harmonized Commodity Description and Coding System (HS) to identify items being shipped. The United States implements the HS in the Harmonized Tariff Schedule of the United States (HTS). Therefore, where the shipment is of foreign origin, it will be identified with HS. Whether HS or HTS, the numbers correlate to the same items. The APHIS chart in the Appendix uses HTS numbers to identify what wood products required a declaration by what date.

For the purposes of the Gibson investigation, two HS codes are relevant. The first is HS 4407— wood split lengthwise that is greater than 6 mm thick. The second is HS 4408—wood veneers, less than 6 mm thick. According to a U.S. customs ruling in 1993, fingerboards for musical instruments, such as guitars and violins, are covered under HS 4407, meaning they are greater than 6 mm thick.[86] Veneer is defined by *Webster's New Collegiate Dictionary* as "a thin sheet of a material: a layer of wood of superior value or excellent grain to be glued to an inferior wood."

According to the affidavit prepared to obtain the Gibson search warrants, "India prohibits the export of products classified under HS Code 4407 for all plant species harvested in India, without exception."[87] Therefore, any wood exported from India under HS 4407 violates Indian law. The affidavit also indicates that exporting wood veneers from India under HS 4408 is legal.[88]

What the Affidavit for the Gibson Warrants States

In order to obtain the search warrants, a Special Agent of FWS completed an affidavit outlining the basis for the search. Four search warrants were ultimately obtained to search Gibson's manufacturing, rough mill, and corporate locations, as well as to search a delivery service. Two shipments formed the basis for the Affidavit: a shipment of ebony from India on June 27, 2011; and a shipment of rosewood and ebony from India on June 20, 2011, which was investigated after the June 27 shipment.[89]

June 27, 2011, Shipment

According to the Affidavit, a CPB official at the Dallas airport suspected a June 27, 2011, shipment of Indian ebony violated the Lacey Act and referred the shipment to an FWS inspector.[90] The FWS inspector, CBP officer, and the APHIS specialist physically identified the shipment as containing sawn Indian ebony of a thickness of 10 mm described under HS 4407.[91] Multiple pieces of documentation accompanied this shipment. (See **Table 1**.) The information on the documents was not consistent, identifying different HS codes (for veneer, for fingerboards, for completed musical instrument parts); and different importers (one in Germany, one in Canada, and Gibson).

Table 1. Documentation Accompanying June 27, 2011 Shipment

Document	Contents as Declared on Form	HS Number	Ultimate Receiver of Goods	Other Information
Air Way Bill 589 2025 5804			Theodor Nagel, Gmbh (of Germany)	Contact Luthier Mercantile upon import. Marked for direct transport to Nashville, TN
Customs and Border Protection Form 3461 (U.S. Customs entry declaration)	Veneer Sheets less than 6 mm	HS 4408.90.0195	Luthier Mercantile (of Canada)	
Indian export declaration (June 10, 2011)	Indian ebony	HS 9209.92.00 (finished parts of musical instruments)		
Certificate of Origin from India	Indian ebony finger boards, 10 mm thick		Nashville Airport	
Lacey Act Declaration (USDA Form PPQ) (June 17, 2011)	Ebony fingerboards for guitars: Diospyros ebenum, harvested in India	HS 4407.99.96	Gibson Guitar Corp.	Document submitted after shipment was referred to FWS for investigation on June 27, 2011

Source: Congressional Research Service, based on information from John M. Rayfield, FWS, *Affidavit in Support of Search Warrant # 11-MJ-1067 A, B, C, D*, pp. 9-11 (August 18, 2011).

Note: Blanks in the table indicate that information was not available, but do not suggest that information was required.

The only form that identified the contents as being under HS 4407 and the ultimate receiver being Gibson, according to the information in the Affidavit, was the Lacey Act declaration submitted by Gibson. The declaration was submitted after the investigation had started. That declaration indicated that the shipment for Gibson contained ebony fingerboards under HS 4407. The problem is that India prohibits export of wood under HS 4407, according to the Affidavit. If the Gibson declaration is true, India export law has been violated. If India export law has been violated, it is possible that the predicate violation of the Lacey Act has been triggered and that a violation of Section

3372(a)(2)(B)(iii) has occurred. As discussed above, that section states that a Lacey Act violation has occurred if an item has been imported, exported, transported, sold, received, acquired, or purchased in interstate or foreign commerce and that item was taken, possessed, transported, or sold in violation of a state or foreign law that governs the export or transshipment of plants. There are no statements in the Affidavit regarding an Indian law protecting ebony or requiring royalties or fees, and thus it appears that FWS did not make a case that there was a violation of either Section 3372(a)(2)(B)(i)—pertaining to foreign laws protecting plants; or Section 3372(a)(2)(B)(ii)—pertaining to foreign laws requiring payment of royalties, taxes, or stumpage fees. Any certification by the Forest Stewardship Council of the timber product would seem irrelevant, as forestry practices do not appear to be the issue in this case.

June 20, 2011, Shipment

The Affidavit also recounts facts relating to an earlier shipment that was discovered after the June 27, 2011, shipment. Again, CPB alerted FWS to a possible Lacey Act violation, in this case Indian rosewood and Indian ebony imported on June 20, 2011.[92] The details about this shipment's documentation are not as clear as for the June 27 shipment. However, the Affidavit indicates that the U.S. customs import declaration identifies Luthier Mercantile as the ultimate receiver of the goods, but that an email from Luthier Mercantile identified Gibson Guitar as the importer.[93] The Affidavit indicates that the shipment was "sawn Indian ebony and sawn Indian rosewood imported under HS code 4407."[94]

While Indian rosewood does not appear to be a protected species under CITES or the ESA, other types of rosewood do have import limitations. CITES lists Brazilian rosewood (*Dalbergia nigra*), Honduran rosewood (*Dalbergia stevensonii*), and Nicaraguan rosewood (*Dalbergia retusa*), thus restricting their trade. No similar restrictions of ebony were found at the time of the Gibson charges, although since then, ebony from Madagascar became listed as an Appendix II species.

Other Shipments to Gibson

The Affidavit indicates that prior to these two shipments, Gibson has submitted Lacey Act declarations for wood products from India 11 times since August 2010.[95] According to the Affidavit, the CPB database shows 11 shipments of rosewood and/or ebony from India to Gibson under HS 4407.[96] FWS brought forfeiture actions for the ebony Gibson imported from Madagascar in 2008 and 2009,[97] and from India in June 2011.[98]

Were Gibson to be charged under Section 3372(a)(B)(iii) for any of these shipments, which seems unlikely in light of the settlement agreement discussed below, rather than DOJ proceeding with a civil forfeiture action only,[99] DOJ would need to show that Gibson *knew* that the shipment violated a law related to export. It would not have to show that Gibson knew the specific Indian or Malagasy law, only that it was doing something illegal.[100] According to the Ninth Circuit, the Lacey Act does not require "that the violator know all the details of the statutes and regulations that make the conduct unlawful. It suffices that he knows, not only that he is importing or exporting [the plant or animal], but also that the [plants or animals] are tainted by a violation of some law."[101] Under such circumstances, a court might consider Gibson's business importing wood;[102] that it had been charged with an import violation before;[103] and/or that the documentation was inconsistent (perhaps to conceal the violation) as evidence that Gibson knew the import was against the law.

Settlement

In July 2012, Gibson reached a settlement with DOJ.[104] The settlement addresses charges and forfeitures from 2008 and 2009 imports, as well as the forfeiture from the 2011 investigation. In exchange for not being prosecuted, Gibson agreed to pay a $300,000 penalty, donate $50,000 to the National Fish and Wildlife Foundation, and withdraw its claims to recover the forfeited wood from Madagascar, reportedly valued at $261,884.[105] The contested wood from India was returned to Gibson.

In the agreement, Gibson acknowledged that it imported wood from Madagascar in 2008 and 2009 after learning that it was illegal. The United States acknowledged that the India tariff rules that were the basis for the 2011 investigation were inconsistent, and agreed not to enforce those rules until India clarified them. Gibson also agreed to enter a Lacey Act compliance program to improve its due care standards for determining wood origins. For example, it agreed to contact foreign governments directly to establish any export restrictions when using a new supplier or source country. It must report its compliance efforts to DOJ two times in the next 18 months. If Gibson complies with these terms, the United States has agreed not to pursue charges.

APPENDIX. DECLARATIONS ENFORCEMENT CHART

I December 15, 2008	II April 1, 2009	III October 1, 2009	IV April 1, 2010
PPS Plant Import Declaration Form will be available on Web site, and accepted after December 15, 2008. Domestic and International Outreach.	HTS Chapters: Ch. 44 Headings (wood & articles of wood) 4401 (Fuel wood) 4403—(Wood in the rough) 4404 (Hoopwood; poles, piles, stakes) 4406—(Railway or tramway sleepers) 4407—(Wood sawn or chipped lengthwise) 4409—(Wood continuously shaped) 4417—(Tools, tool handles, broom handles) 4418—(Builders' joinery and carpentry of wood)	HTS Chapters: Ch. 44 Headings (wood & articles of wood) 4402 Wood charcoal 4412—Plywood, veneered panels, except 44129906 and 44129957 4414—Wooden frames 4419—Tableware & kitchenware of wood 4420 Wood marquetry, caskets, statuettes PLUS PHASE II	HTS Chapters: Ch. 44 Headings (wood & articles of wood) 4421—Other articles of wood Ch. 66 Headings (umbrellas, walking sticks, riding crops) 6602—Walking sticks, whips, crops. Ch. 82 Headings (tools, implements) 8201—Hand tools Ch. 92 Headings (musical instruments) 9201—Pianos 9202—Other stringed instruments Ch. 93 Headings (arms and ammunition) 9302 Revolvers and pistols 93051020—Parts and accessories for revolvers and pistols Ch. 94 Headings (furniture, etc.) 940169—Seats with wood frames Ch. 95 Headings (toys, games & sporting equipment) 950420 Articles and accessories for billiards Ch. 97 Headings (works of art) 9703 Sculptures PLUS PHASES II & III

Source: Congressional Research Service based on data from charts published by APHIS in 74 Fed. Reg. 45415 (September 2, 2009) and 74 Fed. Reg. 5911 (February 3, 2009).

Note: According to APHIS in both of the *Federal Register* notices referenced above, "the failure to submit a declaration will not be prosecuted, and customs clearance will not be denied for lack of a declaration until after the phase-in date above." According to APHIS, furniture other than seats with wooden frames under Ch. 94, Furniture, etc., is removed from the enforcement schedule.

Figure A-1. Phase-In Schedule of Enforcement of the Declaration Requirement for Goods of, or Containing, Plants or Plant Products.

LEGISLATION

While the 113th Congress has not introduced legislation regarding wildlife or plant trafficking, the 112th Congress considered legislation to alter the Lacey Act. Three bills of the 112th Congress proposed limiting Lacey Act enforcement related to imports. H.R. 3210 would have restricted how the Lacey Act applied to wood imported prior to 2008 and composite wood products, and would have allowed an innocent owner defense to forfeiture actions. H.R. 4171/S. 2062 would have eliminated any reference to violations of foreign laws and ended criminal prosecutions for violating the Lacey Act.

End Notes

[1] Act of May 25, 1900, §3, 31 Stat. 187.

[2] Act of May 25, 1900, §2 and §4, respectively, 31 Stat. 188.

[3] 16 U.S.C. §3371(e).

[4] 16 U.S.C. §3371(d).

[5] See 16 U.S.C. §3371(h) (defining Secretary as the Secretary of the Interior and the Secretary of Agriculture); 16 U.S.C. §3373(a) (stating that the provisions will be enforced by "the Secretary, the Secretary of Transportation, or the Secretary of the Treasury").

[6] See Criminal Enforcement Agreement between Department of Justice and Gibson Guitar Corp. (July 27, 2012), http://www.fws.gov/home/feature/2012/USvGibsonGuitarAgreement.pdf.

[7] H.Rept. 110-627; 1981 USCCAN p. 127, 536. (May 13, 2008).

[8] S. Cong. Rec. 15622, at 15635 (December 14, 2007).

[9] S. Cong. Rec. at 13968 (November 6, 2007).

[10] 16 U.S.C. §3371(a).

[11] 16 U.S.C. §3372(a)(1).

[12] The *Examples* in this report should be considered hypothetical situations and not factually or legally true.

[13] 16 U.S.C. §703(a).

[14] 16 U.S.C. §3372(a)(2)(A).

[15] See United States v. McNab, 331 F.3d 1228 (11th Cir. 2003).

[16] United States v. 594,464 Pounds of Salmon, 871 F.2d 824, 826 (9th Cir.1989).

[17] 16 U.S.C. §3372(c).

[18] See United States v. March, 111 Fed. Appx. 888 (9th Cir. 2004).

[19] Fed. R. Civ. P. 44.1: "... In determining foreign law, the court may consider any relevant material or source, including testimony, whether or not submitted by a party or admissible under the Federal Rules of Evidence. The court's determination must be treated as a ruling on a question of law."

[20] United States v. March, 111 Fed. Appx. 888 (9th Cir. 2004).

[21] United States v. One Afghan Urial Ovis Orientalis Blanfordi Fully Mounted Sheep, 964 F.2d 474 (5th Cir. 1992).

[22] United States v. McNab, 331 F.3d 1228, 1232 (11th Cir. 2003).

[23] United States v. McNab, 331 F.3d 1228, 1242 (11th Cir. 2003).

[24] See United States v. Thomas, 887 F.2d 1341, 1348-49 (9th Cir. 1989) (holding that the federal five-year statute of limitations applied, not the state one-year statute of limitations).

[25] United States v. Borden, 10 F.3d 1058, 1062 (4th Cir. 1993) (rejecting the argument that the state one-year statute of limitations applied).

[26] 18 U.S.C. §3282(a). See Agency Holding Corp. v. Malley-Duff & Assocs., Inc., 483 U.S. 143, 155 (1987) (holding in a RICO case that "if no statute specifically defines a limitations period ... for a particular offense, a 'catchall' statute operates to forbid prosecution, trial, or punishment 'unless the indictment is found or the information is instituted within five years next after such offense shall have been committed'").

[27] 16 U.S.C. §3371(f).

[28] Prior to enactment of the 2008 Amendments, the act defined *plant* as: "any wild member of the plant kingdom, including roots, seeds, and other parts thereof (but excluding common food crops and cultivars) which is indigenous to any State and which is either (A) listed on an appendix to the Convention on International Trade in Endangered Species of Wild Fauna and Flora, or (B) listed pursuant to any State law that provides for the conservation of species threatened with extinction." 16 U.S.C. §3371(f) (2007).

[29] An interim final regulatory definition of common food crops and cultivars was issued in July 2013, to take effect September 9, 2013 (definitions of *commercial scale* and *tree*), and August 8, 2013 (all other definitions). 78 Fed. Reg. 40940 (July 9, 2013). A draft definition was issued on August 4, 2010. 75 Fed. Reg. 46859.

[30] 16 U.S.C. §3372(a)(1).

[31] See United States v. Miller, 981 F.2d 439 (9th Cir. 1992) (case pre-dating the 2008 Amendments).

[32] 16 U.S.C. §1538(a)(2).

[33] 16 U.S.C. §1538(a)(2).

[34] 16 U.S.C. §3372(a)(2)(B).

[35] 16 U.S.C. §3372(a)(2)(B)(ii).

[36] See Cong. Rec. E533 (March 13, 2007) (Rep. Blumenauer: "By avoiding export duties, timber royalties and taxes on their profits, companies operating unlawfully are robbing national governments of in excess of $15 billion annually on public lands alone. This loss in revenue decreases governments' ability to invest in the forestry sector to promote sustainable forest management and conserve their natural forest resources").

[37] 16 U.S.C. §3372(a)(2)(B)(iii).

[38] 16 U.S.C. §3372(f)(1).

[39] 74 Fed. Reg. 45415 (September 2, 2009).

[40] 74 Fed. Reg. at 45416 (September 2, 2009).

[41] 16 U.S.C. §3372(f)(4).

[42] 76 Fed. Reg. 10874 (February 28, 2011).

[43] 16 U.S.C. §3372(f)(5).

[44] Personal communication with Abbey L. Fretz, APHIS Congressional Affairs, and the author, July 16, 2013.

[45] P.L. 97-79.

[46] 16 U.S.C. §3373(a)(1). Pursuant to Debt Collection Improvement Act of 1996 (P.L. 104-134), this penalty may be increased by the agency by publishing a notice in the *Federal Register*.

[47] 16 U.S.C. §3373(a)(1).

[48] 16 U.S.C. §3373(a)(1).

[49] S.Rept. 97-123; 1981 U.S.C.C.A.N. 1748, 1757-58.

[50] S.Rept. 97-123; 1981 U.S.C.C.A.N. 1748, 1757.

[51] 16 U.S.C. §3373(d)(2)—but not marking or labeling offenses.

[52] 16 U.S.C. §3373(d)(2).

[53] S.Rept. 97-123; 1981 U.S.C.C.A.N. 1748, 1750.

[54] The Criminal Fines Improvement Act of 1987 enhances criminal fines for those laws whose criminal fines have not been adjusted since 1987. The Criminal Fines Improvement Act applies two steps: first, classification of the offense based on the length of the incarceration provided by the underlying act (18 U.S.C. §3559); and second, calculation of the fine based on the offense classification (18 U.S.C. §3571).

[55] 16 U.S.C. §3373(d)(1).

[56] 16 U.S.C. §3373(d)(1)(A).

[57] 16 U.S.C. §3373(d)(1)(B).

[58] S.Rept. 97-123; 1981 U.S.C.C.A.N. 1748, 1758.

[59] S.Rept. 97-123; 1981 U.S.C.C.A.N. 1748, 1759.

[60] 16 U.S.C. §3373(d)(1). The Lacey Act indicates a maximum fine of $20,000. The fines in the text are based on the enhanced amounts under the Criminal Fines Improvement Act.

[61] For a broader analysis of forfeiture, see CRS Report 97-139, *Crime and Forfeiture*, by Charles Doyle.

[62] United States v. 144,774 Pounds of Blue King Crab, 410 F.3d 1131 (9th Cir. 2005) (holding that crab taken illegally was property that was illegal to possess and therefore fit the definition of contraband).

[63] United States v. One Afghan Urial Ovis Orientalis Blanfordi Fully Mounted Sheep, 964 F.2d 474 (5th Cir. 1992).

[64] One Lot Emerald Cut Stones v. United States, 409 U.S. 232, 237 (1972).

[65] 16 U.S.C. §3374(a)(1).

[66] 16 U.S.C. §3374(a)(2).

[67] 16 U.S.C. §3374(a)(1) (emphasis added).

[68] S.Rept. 97-123; 1981 U.S.C.C.A.N. 1748, 1760.

[69] S.Rept. 97-123; 1981 U.S.C.C.A.N. 1748, 1760.

[70] United States v. One Afghan Urial Ovis Orientalis Blanfordi Fully Mounted Sheep, 964 F.2d 474, 476 (5th Cir. 1992). See also United States v. 2,507 Live Canary Winged Parakeets, 689 F. Supp. 1106, 1117 (S.D. Fla. 1988).

[71] United States v. Proceeds from Sale of Approximately 15,538 Panulirus Argus Lobster Tails, 834 F. Supp. 385 (S.D. Fla. 1993).

[72] United States v. 144,774 Pounds of Blue King Crab, 410 F.3d 1131 (9th Cir. 2005).

[73] 16 U.S.C. §3374(d).

[74] P.L. 106-185.

[75] 18 U.S.C. §983(c)

[76] 18 U.S.C. §983(f).

[77] 18 U.S.C. §983(d).

[78] 18 U.S.C. §983(d)(4).

[79] S.Rept. 97-123; 1981 U.S.C.C.A.N. 1748, 1760.

[80] H.Rept. 110-882.

[81] U.S. Department of the Interior v. Three Pallets of Tropical Hardwood (Crouch), INV. No. 2009403072, at 3 (Office of the DOI Solicitor June 22, 2010).

[82] U.S. Department of the Interior v. Three Pallets of Tropical Hardwood (Crouch), INV. No. 2009403072, at 4 (Office of the DOI Solicitor June 22, 2010). The Solicitor found no mitigating circumstances "[the importer] did not do all he could within his power to comply

with regulations and ensure that the shipment was authorized by an export permit that properly documented the required information." *Id.* at 6.

[83] The Forest Stewardship Council (FSC) is a non-profit group that reviews forest management and production practices. It describes its certification as providing a basis "for socially and environmentally responsible purchasing decisions." See FSC website, http://www.fsc.org/about-fsc.html.

[84] *Gibson Guitar Corp. Responds to Federal Raid* (August 25, 2011), available at http://www.gibson.com/en-us/ Lifestyle/News/gibson-0825-2011.

[85] Criminal Enforcement Agreement between Department of Justice and Gibson Guitar Corp. (July 27, 2012), http://www.fws.gov/home/feature/2012/USvGibsonGuitarAgreement.pdf.

[86] Tariff classification ruling NY 881630 (January 26, 1993) (pertaining to sawn Indian rosewood (*Dalbergia latifolia*) and sawn Indian ebony (*Diospyros ebenum*)).

[87] John M. Rayfield, FWS, *Affidavit in Support of Search Warrant # 11-MJ-1067 A, B, C, D*, p. 4 (August 18, 2011) (hereinafter Affidavit). See also Affidavit, p. 8 ("Further research of ... all other published laws and regulations of the Government of India ... revealed no exceptions to this export prohibition").

[88] Affidavit at 7.

[89] Neither Indian rosewood (*Dalbergia latifolia*) nor ebony (*Diospyros ebenum*) appears to be species protected under the Convention on International Trade of Endangered Species of Wild Fauna and Flora (CITES) or other American law at the time of the violation. Ebony from Madagascar, however, is a CITES Appendix II species as of June 12, 2013.

[90] Affidavit at 8.

[91] Affidavit at 8, 11.

[92] Affidavit at 14 (stating the shipment was discovered on July 28, 2011).

[93] Affidavit at 14.

[94] Affidavit at 14.

[95] Affidavit at 11-12.

[96] Affidavit at 12-13.

[97] United States v. Ebony Wood in Various Forms, No. 3:10-cv-747 (M.D. Tenn. *administratively closed* September 28, 2011). See Heath E. Combs, *Gibson Raid Raises Attention on Lacey Act Enforcement*, Furniture Today (September 6, 2011), available at http://www.furniture today.com/article/print/542690-Gibson_raid_raises_attention_on_Lacey_Act_ enforcement. php.

[98] United States v. 25 Bundles of Indian Ebony Wood, No. 3:11-cv-00913 (M.D. Tenn.).

[99] Forfeiture action has been initiated. United States v. Indian Ebony and Rosewood, No. 3:12-mc-00014 (M.D. Tenn. 2011) (complaint not yet filed).

[100] See, e.g., United States v. Santillan, 243 F.3d 1125 (9th Cir. 2001) (holding that culpability was demonstrated by the fact that he lied about the parrots he was hiding in his car while crossing into the United States from Mexico).

[101] United States v. Santillan, 243 F.3d 1125, 1129 (9th Cir. 2001).

[102] See United States v. Three Pallets of Tropical Hardwood (Crouch), INV. No. 2009403072, at 5 (Office of the DOI Solicitor June 22, 2010) (rejecting a defense that the declaration requirement was new, holding that "a company that specializes in international shipments should be well versed in the applicable laws that govern such transactions").

[103] See United States v. Kuipers, 49 F.3d 1254 (7th Cir. 1995) (allowing DOJ to introduce other crimes in a Lacey Act case to show defendant *knew* he was violating the law).

[104] Criminal Enforcement Agreement between Department of Justice and Gibson Guitar Corp. (July 27, 2012), http://www.fws.gov/home/feature/2012/USvGibsonGuitarAgreement.pdf.

[105] Department of Justice Press Release, *Gibson Guitar Corp. Agrees to Resolve Investigation into Lacey Act Violations* (August 6, 2012), available at http://www.justice.gov/opa/pr/2012/August/12-enrd-976.html.

In: Climate, Energy, and Environment. Vol. 1 ISBN: 978-1-62948-407-5
Editor: Jake S. Diaz

Chapter 4

ENERGY AND ECONOMIC IMPACTS OF U.S. FEDERAL ENERGY AND WATER CONSERVATION STANDARDS ADOPTED FROM 1987 THROUGH 2011*

Stephen Meyers, Alison Williams and Peter Chan

ABSTRACT

This paper presents estimates of the key impacts of Federal energy and water conservation standards adopted from 1987 through 2011. The standards for consumer products and commercial and industrial equipment include those set by legislation as well as standards adopted by DOE through rulemaking.

In 2011, the standards saved an estimated 3.2 quads of primary energy, which is equivalent to 3% of total U.S. energy consumption. The savings in operating costs for households and businesses totaled $42 billion. The average household saved $309 in operating costs as a result of residential and plumbing product standards. The estimated reduction in CO2 emissions associated with the standards in 2011 was 176 million metric tons, which is equivalent to 3 percent of total U.S. CO2 emissions.

The estimated cumulative energy savings over the period 1990-2070 amount to 176 quads. Accounting for the increased upfront costs of more-

* This is an edited, reformatted and augmented version of a Lawrence Berkley National Laboratory publication, report number LBNL-5515E, dated April 2012.

efficient products and the operating cost (energy and water) savings over the products' lifetime, the standards have a projected cumulative net present value (NPV) of consumer benefit of between $938 billion and $1,279 billion, using 7 percent and 3 percent discount rates, respectively.

The water conservation standards, together with energy conservation standards that also save water, reduced water use by 1.5 trillion gallons in 2010, and will achieve cumulative water savings by 2040 of 51.4 trillion gallons. The estimated consumer savings in 2010 from reduced water use amounted to $10.8 billon.

Standards set by legislation account for 47 percent of total cumulative primary energy savings and 55 to 66 percent of the total cumulative NPV (depending on which discount rate is used). Standards set by DOE rulemakings account for 53 percent of total cumulative primary energy savings and 34 to 45 percent of the total cumulative NPV.

INTRODUCTION

The energy conservation program for consumer products and certain commercial and industrial products was established by the Energy Policy and Conservation Act of 1975 (EPCA). EPCA established a program consisting of test procedures, labeling, and energy conservation targets for 19 types of consumer products. The National Energy Conservation Policy Act of 1978 amended EPCA by replacing the energy conservation targets program and directing that energy conservation standards be set for the covered consumer products. With the passage of the National Appliance Energy Conservation Act (NAECA) in 1987, EPCA was further amended to establish the first national energy conservation standards for consumer products. Subsequent amendments in 1988, 1992, 2005, and 2007 further expanded the scope of coverage to include additional consumer products, certain commercial and industrial equipment, as well as water conservation standards for residential and commercial products. EPCA, as amended, requires the Department of Energy (DOE) to update or establish standards at levels that "achieve the maximum improvement in energy [or water] efficiency ... which the Secretary determines is technologically feasible and economically justified." EPCA defines "economically justified" standards as those for which benefits exceed the costs, given a number of factors, including impacts on consumers and manufacturers and the nation's need to save energy or water.

This paper presents estimates of the key impacts of the energy and water conservation standards that have been adopted from 1987 through 2011. It updates the results presented in Meyers et al. (2011), which covered standards adopted through 2010. The standards covered include those set by legislation as well as standards adopted by DOE through rulemaking. The estimates cover both historic and projected impacts of these standards. The impacts cover primary energy savings and water savings, net present value of consumer[1] benefits, and estimated reductions in CO_2 emissions.

Table 1 lists products covered by standards, the year(s) compliance was or will be required, and the legislation that initially authorized each standard. The standards that were adopted in 2011 cover the following products (compliance year in parentheses):

- Residential furnaces (2013)
- Residential refrigerator-freezers (2014)
- Residential freezers (2014)
- Room air conditioners (2014)
- Residential clothes dryers (2014)
- Fluorescent lamp ballasts (2014)
- Central air conditioners and heat pumps (2015)

Table 1. Federal Energy and Water Conservation Standards for Appliances and Equipment Adopted From 1987 Through 2011

Product	Compliance Date for Original Standard and Updates	Authorizing Legislation*
RESIDENTIAL		
Clothes Washers+	1988, 1994, 2004/2007	NAECA 1987
Clothes Dryers	1988, 1994, 2014	NAECA 1987
Dishwashers +	1988, 1994, 2010	NAECA 1987
Refrigerators and Refrigerator-Freezers	1990, 1993, 2001, 2014	NAECA 1987
Freezers	1990, 1993, 2001, 2014	NAECA 1987
Room Air Conditioners	1990, 2000, 2014	NAECA 1987
Central Air Conditioners and Heat Pumps	1992/1993, 2006, 2015	NAECA 1987
Water Heaters	1990, 2004, 2015	NAECA 1987
Furnaces	1992, 2013	NAECA 1987
Boilers	1992, 2012	NAECA 1987
Direct Heating Equipment	1990, 2013	NAECA 1987
Cooking Products	1990, 2012	NAECA 1987
Pool Heaters	1990, 2013	NAECA 1987
Ceiling Fans and Ceiling Fan Light Kits	2007	EPACT 2005
Torchieres	2006	EPACT 2005

Table 1. (Continued)

Product	Compliance Date for Original Standard and Updates	Authorizing Legislation*
Dehumidifiers	2007, 2012	EPACT 2005
External Power Supplies	2008	EISA 2007
COMMERCIAL & INDUSTRIAL		
Electric Motors	1997, 2010	EPACT 1992
Warm Air Furnaces	1994	EPACT 1992
Packaged Boilers	1994	EPACT 1992
Air Conditioners and Heat Pumps	1994/1995, 2003/2004, 2010, 2012	EPACT 1992
Water Heaters, Hot Water Supply Boilers and Unfired Hot Water Storage Tanks	1994, 2004	EPACT 1992
Distribution Transformers	2007, 2010	EPACT 1992, EPACT 2005
Refrigerators, Refrigerator-Freezers and Freezers	2010, 2012	EPACT 2005
Automatic Ice Makers	2010	EPACT 2005
Clothes Washers+	2007	EPACT 2005
Unit Heaters	2008	EPACT 2005
Refrigerated Beverage Vending Machines	2012	EPACT 2005
Walk-in Coolers and Walk-in Freezers	2009	EISA 2007
LIGHTING PRODUCTS		
Fluorescent Lamp Ballasts	1990, 2005/2010, 2014	NAECA 1988
General Service Fluorescent Lamps and Incandescent Reflector Lamps	1995, 2008, 2012	EPACT 1992, EISA 2007
Medium Base Compact Fluorescent Lamps	2006	EPACT 2005
Illuminated Exit Signs	2006	EPACT 2005
Traffic Signal Modules and Pedestrian Modules	2006	EPACT 2005
Mercury Vapor Lamp Ballasts	2008	EPACT 2005
Metal Halide Lamp Ballasts and Fixtures	2009	EISA 2007
General Service Incandescent Lamps, Intermediate Base Incandescent Lamps and Candelabra Base Incandescent Lamps	2012/2014 & 2020	EISA 2007
PLUMBING PRODUCTS		
Faucets++	1994	EPACT 1992
Showerheads++	1994	EPACT 1992
Water Closets++	1994/1997	EPACT 1992
Urinals++	1994/1997	EPACT 1992

Product	Compliance Date for Original Standard and Updates	Authorizing Legislation*
Pre-rinse Spray Valves++	2007	EPACT 2005

* The Energy Policy and Conservation Act of 1975 was amended to set energy or water conservation standards by the National Appliance Energy Conservation Act (NAECA 1987), the National Appliance Energy Conservation Amendments of 1988 (NAECA 1988), the Energy Policy Act of 1992 (EPACT 1992), the Energy Policy Act of 2005 (EPACT 2005), and the Energy Independence and Security Act of 2007 (EISA 2007).

+ Water and energy conservation standard

++ Water conservation standard

Analysis Method Overview

Different analytical methods were used for five sets of standards. For NAECA 1987 and NAECA 1988 standards and DOE updates of those standards issued before 2007, we utilized the analyses conducted by Lawrence Berkeley National Laboratory (LBNL) in 2007-2008.[1] For EPACT 1992 standards, we developed new estimates for this study. For EPACT 2005 standards, we reviewed and utilized an analysis conducted by Nadel et al.[2] and added information from DOE analyses where available. For most of the EISA 2007 standards, we drew upon an analysis conducted by DOE.[3] For the other EISA 2007 standards,[2] we used unpublished national impact analyses that were prepared by LBNL for DOE. For standards adopted by DOE in 2007-2011, we drew on the national impact analyses performed for the rulemakings for each of the standards and adapted the results for the framework of this study. Appendix A further describes the use of the above sources in this study.

It is important to note that the analyses performed for the rulemakings for each of the standards were highly detailed and were carefully reviewed by stakeholders. All of the other sources used for this study were much less detailed in their approach and less extensively reviewed.

The most challenging aspect of estimating the impacts of standards is characterizing what would have happened without new or amended standards. We call this counterfactual against which impacts of standards is measured the "base case." The sources used for this study vary in how they characterized the base case. The LBNL analysis of the NAECA standards and DOE updates of those standards before 2007 estimated a dynamic base case in which the energy efficiency of the products improves somewhat even without standards. The analyses performed for DOE's rulemakings also generally consider how the market might change in the absence of new or amended standards. In contrast, the analyses used for EPACT 1992, EPACT 2005, and EISA 2007 standards used simple assumptions (in many cases, no change in efficiency) regarding the base case.

We focused on three key impacts associated with standards: (1) primary energy savings; (2) additional installed costs; and (3) operating cost savings. Operating cost savings primarily consist of energy cost savings. For standards that save water, we also included water cost savings where possible. In some cases (primarily the DOE rulemakings in 2007-2011), the operating cost savings also include changes in maintenance and repair costs associated with the standards.

From (2) and (3) we derived net cost or savings for each year. We also estimated reductions in CO_2 emissions associated with the standards using the annual primary energy savings and annual average CO_2 emissions factors for the electricity generation sector and for natural gas.

For each standard we developed a time series of annual impacts, with economic impacts expressed in 2010 dollars. For the NAECA standards and DOE updates of those standards before 2007, and for standards adopted by DOE in 2007-2011, we estimated annual impacts for each standard for 30 years worth of shipments.

For most of the other standards, for which the base case often assumes no change in efficiency, we used a shorter period of shipments as a way of compensating for the lack of a dynamic base case, which might tend to overstate the savings from standards. For all standards, we estimated annual energy savings and operating cost savings until products installed in the final year considered are retired from the stock. Retirement is based on the average lifetime for each product.

Using the annual economic impacts, we derived a net present value (NPV) by discounting future impacts to the present (defined as 2010 for purposes of analysis).

For economic impacts occurring after 2010, we used discount rates of 3% and 7%, which are the rates used by DOE in its rulemaking analyses, in accordance with guidance from the Office of Management and Budget to Federal agencies on the development of regulatory analysis.[4] For economic impacts occurring before 2010, we derived estimates of their present value using interest rates of 3% and 7%. This approach reflects the view that the present value of the past stream of benefits should reflect the returns to those "profits" had they been invested elsewhere in the economy.

We estimated a monetary value of the reductions in CO_2 emissions using the mid-range series for the global Social Cost of Carbon (SCC) developed by a Federal interagency process.[3] The series used has a value of $22.3 per metric ton in 2010 and increases at 3 percent per year.

Product-Level Impact of Standards: The Case of Refrigerator-Freezers

Figure 1 illustrates how standards have had an important effect on the energy efficiency of new products, in this case refrigerator-freezers. The average new refrigerator-freezer in 2010 used only 44% of the energy per year as an average new unit in 1985.

Total energy use for these products has declined even as shipments increased and the average size of new units grew. Nationally, in 2010 refrigerator-freezers used one-third less total energy than in 1985 even though there were 70 million more units in use.[4]

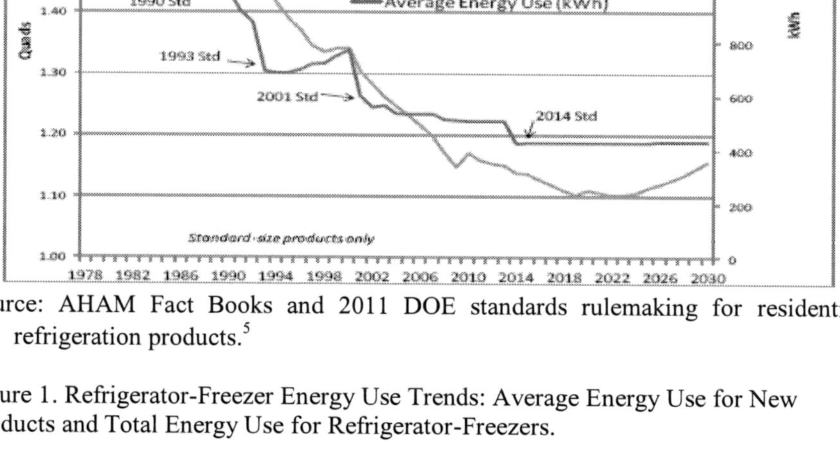

Source: AHAM Fact Books and 2011 DOE standards rulemaking for residential refrigeration products.[5]

Figure 1. Refrigerator-Freezer Energy Use Trends: Average Energy Use for New Products and Total Energy Use for Refrigerator-Freezers.

NATIONAL IMPACTS[6]

In 2010, the energy and water conservation standards saved an esti mated 2.9 quads of primary energy, which is equivalent to 3% of total U U.S. energy consumption.

The savings in operating costs totaled $40 billion.

As shown in Table 2, the cumulative primary energy savings through 2010 amount to 25.9 quads. The cumulative consumer NPV through 2010, which equals the operating cost savings minus the additional product costs associated with standards, is between $275 billion and $348 billion.

Over the enti re ti me period considered (1990-2070), the cumulative primary energy savings amount to 176 quads and the cumulative consumer N PV is between $938 billion and $1,279 billion (Table 3). Residential products account for more than half of the total cumulative primary energy savings. In addition to energy cost savings from energy conservation standards, the consumer NPV includes water cost savings from those standards that affect both energy and water use (such as standards on clothes washers), as well as energy cost savings from water conservation standards that save hot water (i.e., standards on faucets and showerheads).

The cumulative energy savings achieved through 2010 are only 15 percent of the total cumulative energy savings. Thus, most of the savings from standards already adopted will occur in the future. Furthermore, future savings will grow as new standards are adopted.

Table 2. Cumulative Primary Energy Savings and Cumulative Consumer NPV Through 2010 for all Standards

Product type	Primary energy savings (quads)	Share of primary energy savings	Cumulative NPV (billion 2011$)*	Share of NPV
Residential	15.1	58%	$98 - $124	35-36%
Commercial & Industrial	3.0	12%	$17 - $20	6%
Lighting Products	5.0	19%	$47 - $63	17-18%
Plumbing Products	2.8	11%	$114 - $142	41%
Total	25.9	100%	$275 - $348	100%

* The lower value refers to the NPV using 7% discount rate, while the higher value refers to the NPV using 3% discount rate.

Table 3. Cumulative Primary Energy Savings and Consumer NPV for All Standards (1990- 2070

Product type	Cumulative primary energy savings (quads)	Share of total cumulative primary energy savings	Cumulative NPV (billion 2011$)*	Share of total cumulative NPV
Residential	96.1	55%	$437 - $654	47-51%
Commercial & Industrial	28.1	16%	$71 - $115	8-9%
Lighting Products	43.2	25%	$169 - $241	18-19%
Plumbing Products	8.3	5%	$260 - $269	21-28%
Total	175.7	100%	$938 - $1,279	100%

* The lower val ue refers to the NPV using 7% di scount rate, while the higher val ue refers to the NPV using 3% di scount rate.

As shown in Table 4, standards set by legislation account for 72 percent of cumulative energy savings through 2010, and DOE rulemakings account for 28 percent. Standards set by legislation account for 47 percent of total cumulative primary energy savings and 55-66% of the total consumer NPV (depending on which discount rate is used). Standards set by DOE rulemakings account for 53 percent of total cumulative primary energy savings and 34-45% of the total consumer NPV (see Table 5).

Table 4. Cumulative Primary Energy Savings and Consumer NPV Through 2010 for All Standards by Source of the Standards

Source of Standards	Cumulative Primary Energy Savings (quads)	Consumer NPV at 3% discount rate (billion 2011$)	Consumer NPV at 7% discount rate (billion 2011$)
Legislation			
NAECA 1987/1988	10.3	95.3	127
EPACT 1992	7.3	149	186
EPACT 2005*	0.6	5.0	5.3
EISA 2007*	0.4	-3.0	-3.1
Subtotal	18.6	246	315
DOE Rulemakings			
Subtotal	7.3	29.0	33.3
Total	25.9	269.6	341.1

* Savings through 2010 are very small since the standards are just taking effect.

Table 5. Cumulative Primary Energy Savings and NPV for all Standards by Source of the Standards (1990-2070

Source of Standards	Primary Energy Savings (quads)	Consumer NPV at 3% discount rate (billion 2011$)	Consumer NPV at 7% discount rate (billion 2011$)
Legislation			
NAECA 1987/1988	21.3	170	181
EPACT 1992	15.3	317	314
EPACT 2005	15.5	101	62.5
EISA 2007	30.7	116	60.4
Subtotal	82.8	704	618
DOE Rulemakings			
Subtotal	92.9	576	320
Total	175.7	1279	938

Table 6 presents the annual and cumulative water savings from standards. The results include water savings from water conservation standards as well as from energy conservation standards that also save water (such as standards on clothes washers and dishwashers).[7] The annual savings of 1.5 trillion gallons in 2010 are equal to 9 percent of the total water withdrawals for public supply in 2005.[8] The estimated dollar savings from reduced water use in 2010 amounted to $10.8 billon. Lower water use in 2010 due to standards also reduced energy use in municipal sewage and water systems by an estimated 0.05 quad.[9]

Table 6. Annual and Cumulative Water Savings for All Water-Conserving Standards

	Water Savings (trillion gallons)	
	Annual	Cumulative through
2010	1.5	11.7
2025	1.5	38.6
2040	0.3	51.4

As shown in Table 7, the estimated reduction in CO2 emissions associated with the standards in 2010 was 167 million metric tons of CO_2, which amounts to 3.0% of total U.S. CO_2 emissions in 2010. The estimated economic (present) value of reductions in CO2 emissions associated with the standards through 2055 is $194 billion.

Table 7. Annual and Cumulative Reduction in Carbon Dioxide Emissions for All Energy Conservation Standards

	CO_2 Emissions Reduction (million tons CO_2)		Present Value (billion 2011$)*	
	Annual	Cumulative through	Annual	Cumulative through
2010	167	1544	3.8	37
2025	280	5299	5.7	117
2040	132	8655	2.3	181
2055	20	9529	0.3	194

* The present value was calculated using a discount rate of 3%, in keeping with the method used by DOE in recent rulemakings. See discussion in appendix A.

Figure 2 shows the annual primary energy savings for each grouping of standards, and Figure 3 shows the annual undiscounted net consumer impact. The impacts will peak in the 2020-2025 period as purchases of products subject to standards increase. The decline in impacts reflects the analytical convention used to count impacts for 25-30 years of shipments for each standard. The annual savings peak at around $48 billion in the 2020-2025 period.

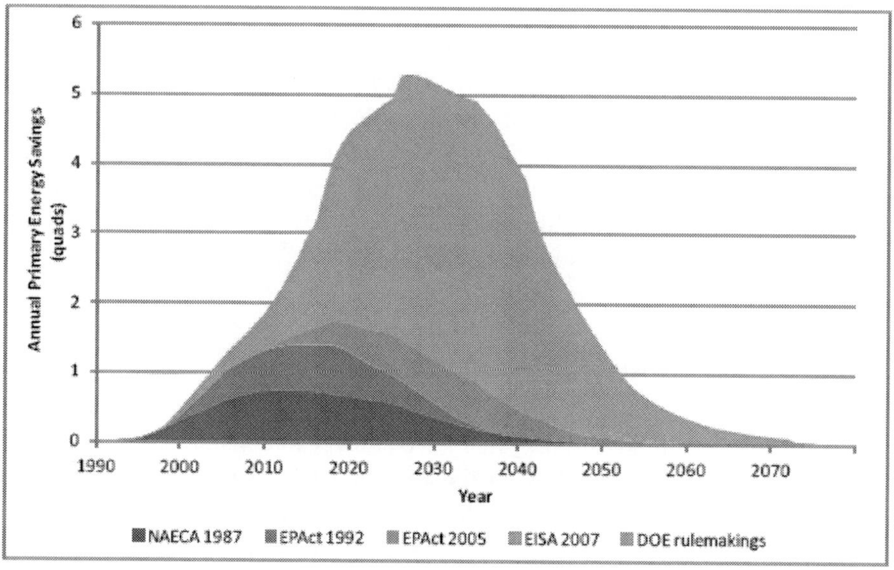

Figure 2. Annual Primary Energy Savings for all Standards by Source of the Standards.

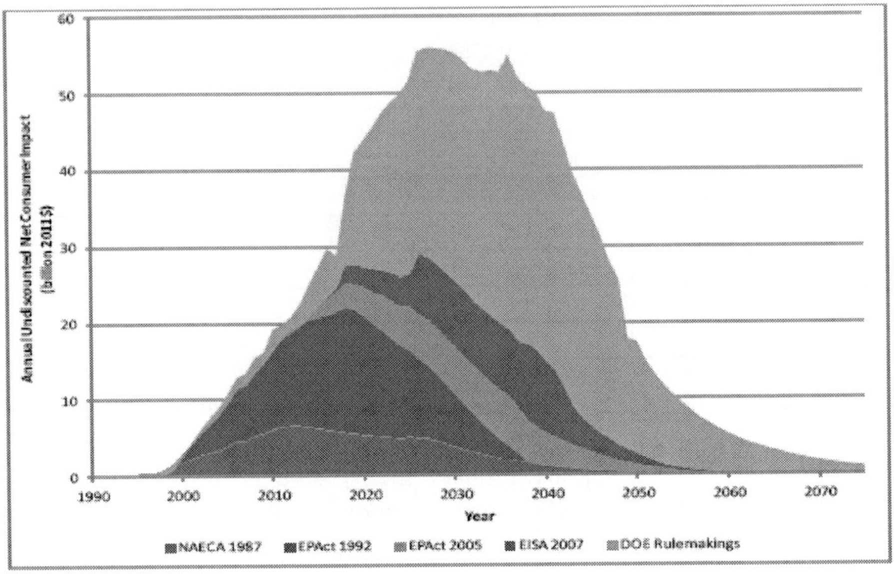

Figure 3. Annual Undiscounted Net Consumer Benefit for all Standards.

Consumer Impacts

In 2010, we estimate that the typical household saved $289 in energy and water expenses as a result of residential and plumbing product standards. By 2010, most U.S. households used one or more appliances that were subject to Federal energy or water conservation standards. On average, the primary energy savings from residential and plumbing standards in 2010 amounted to 19 million Btu per household, which is equivalent to 10 percent of the total energy use of 186 million Btu per household.[10]

Sources of Uncertainty

The estimates made by this study are subject to considerable uncertainty. A major source of uncertainty is the assumed base case against which the impacts of standards are measured. In principle, a base case should reflect best judgment as to how the market for a given product will evolve without the standards under consideration. Estimating the consumer demand for higher efficiency products is difficult. Even more difficult is estimating what other policies, either Federal or State, might be implemented if there were no Federal efficiency standards.

We have more confidence in the estimates of per-unit energy savings and additional cost in the sources used for this study, though these too are subject to uncertainty. Most of the sources assume that the incremental costs of higher efficiency remain constant over time.[11] This assumption likely overstates the true costs for two reasons. First, manufacturers of appliances and equipment often find ways to reduce the cost of producing higher efficiency products. Second, inflation-adjusted prices of many types of appliances and equipment have trended downward in recent decades. To the extent that this trend continues, it means that the incremental cost of higher efficiency products may decline over time.

The estimates of primary energy savings in the sources we used are based on estimates of "site" energy savings (i.e., savings where the product is in operation). Most of the sources convert site savings to primary savings using an average multiplier. In contrast, the National Impact Analysis spreadsheets from the DOE rulemakings incorporate marginal site-to-source energy conversion factors. These factors represent the response of the electricity system to an incremental decrease in consumption associated with appliance standards. DOE uses annual site-to-source conversion factors based on a version of the Energy Information Administration's National Energy Modeling System (NEMS). The marginal factors are lower than average site-to-source conversion factors and are likely more accurate. If we had been able to apply marginal site-tosource conversion factors to all of the standards included, the estimated primary energy savings (and also the reductions in CO_2 emissions) would be lower.

For consumer cost savings that occurred in the past, there is some question as to whether the compounding of past savings used in this study is appropriate. We have not found clear guidance in the literature, but there is some precedent for the practice of compounding past savings to estimate their present value.[12] There is uncertainty regarding the extent to which the savings from appliance standards were invested elsewhere in the economy, and what the appropriate interest rate should be. Without compounding, the cumulative consumer NPV for all standards through 2010 would be around 15 percent less than reported here.

There is some evidence that consumers use higher efficiency appliances more intensively due to the reduction in operating cost. The extent of this so-called direct rebound effect varies among products.[5] In recent years DOE has accounted for a rebound effect in many of its rulemakings. Thus, the energy savings estimates for many standards adopted by DOE since 2008 include an adjustment (subtraction) for a rebound effect.[13] The other sources used for this study do not include such an adjustment. The lack of this adjustment means that the savings may be overestimated by 5%-10%.

CONCLUSION

We estimate that energy efficiency standards for residential appliances and commercial equipment adopted from 1987 through 2011 have saved a total of 29 quads through 2011, an amount equal to one-fourth of total annual U.S. energy use. In 2011, the standards saved an estimated 3.2 quads of primary energy, which is equivalent to 3% of total U.S. energy consumption. The savings in operating costs for households and businesses totaled $42 billion. The average household saved $309 in operating costs as a result of residential and plumbing product standards. The estimated reduction in CO_2 emissions associated with the standards in 2011 was 176 million metric tons of CO2, which is equivalent to 3 percent of total U.S. CO_2 emissions.

The majority of the savings attributable to the standards adopted thus far are still to come, as products subject to the standards enter the stock. The standards are projected to achieve cumulative energy savings of 176 quads. Accounting for the increased upfront costs of more-efficient products and the operating cost (energy and water) savings over the products' lifetime, the standards have a projected cumulative net present value (NPV) of consumer benefit of between $938 billion and $1,279 billion, using 7 percent and 3 percent discount rates, respectively.

The water conservation standards, together with energy conservation standards that also save water, reduced water use by 1.5 trillion gallons in 2010, and will achieve cumulative water savings by 2040 of 51.4 trillion gallons. The estimated consumer savings in 2010 from reduced water use amounted to $10.8 billon.

Standards set by legislation account for 47 percent of total cumulative primary energy savings and 55 to 66 percent of the total cumulative NPV (depending on which discount rate is used). Standards set by DOE rulemakings account for 53 percent of total cumulative primary energy savings and 34 to 45 percent of the total cumulative NPV.

APPENDIX A: METHODS FOR ESTIMATING NATIONAL IMPACTS FROM STANDARDS[14]

NAECA 1987 and 1988 and DOE Updates before 2007

For all of the standards except one, we used the data developed by Meyers *et al.*[1] That study developed a spreadsheet accounting model to calculate energy savings and consumer costs and savings for each product. The model tracks the energy use of products sold in each year, beginning in the late 1980s. The model uses historic and projected data on annual shipments of each product and subtracts units from the stock using a retirement function based on the estimated average lifetime of each product.

The key feature of the model is that it associates a specific average energy consumption and average product price for each vintage of a given product. (A vintage refers to the products shipped in a given year.) Both of these variables are a function of the energy efficiency assigned to each vintage. In most cases, the actual energy efficiency for each vintage of a product is assigned based on industry sources.

The approach for estimating the impacts of standards involves deriving a base case scenario for average energy efficiency and product price that assumes no standards were or will be implemented. In principle, the base case assumes energy efficiency increases over time as a result of all factors that shape energy efficiency other than Federal standards. For further discussion, see section 2 of Meyers *et al.*

For the commercial heating, air conditioning, and water heating standards with compliance dates of 2003 and 2004, we started from the following data reported by Belzer and Winiarski:[6] (1) primary energy savings cumulative through 2030 and (2) net economic impacts at a 7-percent discount rate cumulative from units shipped through 2030. We used an average lifetime for these products of 15 years.

We assume that units retire uniformly over the lifetime and that the annual energy savings will go up after the effective date until it stabilizes when all the pre-standard units have been replaced by units meeting the standards. This period that it takes for the annual energy savings to reach its maximum is equal to the lifetime of the product. Using these assumptions, we calculate the annual site and primary energy savings that will match the given cumulative energy savings from 2003 to 2030. Then we used the Excel Solver to solve for the unit energy saving and incremental equipment cost per unit that will give a net present value (NPV) that closely matches the given NPV at a 7-percent discount rate. We then extended the time series to include shipments through 2032 to yield a 30-year analysis period.

EPACT 1992

We developed new estimates for this study, as described below. We assumed no change in base case efficiency over time. To compensate for potential overstatement of savings, we counted impacts for only 20 years worth of product shipments. Further details may be found in spreadsheets that are available from the authors.

Commercial Furnaces and Boilers, Air Conditioners and Heat Pumps, and Water Heaters

We modified the analytical structure and some of the data developed by Rosenquist et al. for the 2004 study for the National Commission on Energy Policy (NCEP).[7]

We estimated base case efficiencies and unit incremental costs for these products using PNNL (2000). This report presents average efficiencies in 1999 and costs for both an EPACT 1992 baseline product and an average product in 1999. We applied these differentials to derive an approximate pre-EPACT 1992 baseline efficiency and contractor cost for each product.

Electric Motors

We developed a simplified NIA model to estimate the impacts of the EPACT 1992 standards for electric motors, using one "average motor" as the basis for the calculations.

The "average motor" energy use was calculated in the base case and in the standards case, using market-weighted averages across the covered horsepower (hp) ranges, pole configurations, and enclosure type to determine the following parameters: operating hours, load, lifetime, horsepower, and efficiency. All inputs were derived from the draft preliminary analysis from DOE's 2011 rulemaking for electric motors. The base-case efficiency is estimated assuming 30% of shipped motors are at pre-EPACT standard efficiency levels, 30% are already at the EPACT 1992 efficiency levels, and 40% are at National Electrical Manufacturers Association (NEMA) premium efficiency levels. The standards-case efficiency is estimated using a "roll-up" scenario, which leads to assuming 60% of motors are at the EPACT 1992 efficiency levels and 40% are at the NEMA premium efficiency levels. Motor equipment costs (includes the repair costs) for the "average motor" in the base case and standards case were estimated by extrapolating price and weight data from the preliminary analysis. Repairs are assumed to occur after 5 years of usage and once in a motor's lifetime.

Shipment data were obtained from the preliminary analysis and are assumed to be the same in the base-case and in the standards-case. The market-weighted average lifetime (12 years) was used to calculate the affected stock.

National site energy savings were obtained from multiplying the affected stock by the difference in energy use between the base case and standards case for the "average motor". National equipment incremental costs were calculated using the affected stock multiplied by the difference in equipment costs between the base case and standards case for the "average motor".

Fluorescent Lamps and Incandescent Reflector Lamps

Fluorescent Lamps

We calculated savings for full-wattage T12 lamps covered by the standards sold after the effective dates of the standards: April 30, 1994 for 8-foot T12 and 8-foot T12/HO lamps and October 31, 1995 for 4-foot lamps. To calculate fluorescent lamp shipments, we adapted the spreadsheet used to analyze the impacts of the NAECA fluorescent ballast standards by Meyers *et al.* The base-case forecast assumed that 60 percent of lamp shipments in 1994 were full-wattage lamps, while 40 percent were reduced-wattage lamps already complying with the EPAct 1992 standards, according to a 1989 report on Massachusetts' lamp standards by Nadel et al.[8] Since the lamps covered by the EPAct 1992 lamp standards ("covered lamps") were used with magnetic ballasts, with very few T12 lamps used with electronic ballasts, we assumed that lamp shipments tracked the pattern of magnetic ballast shipments. When the fluorescent ballast standards came into effect in 2005 for ballasts in new luminaires, there was a corresponding substantial decrease in T12 lamp shipments. By 2010, when the ballast standards took effect for the renovation market as well, very few T12 lamps were sold (in the commercial/industrial market).The shipments of covered fluorescent lamps for 1994 were based on estimates by Geller and Nadel.[9] For 1995 - 2010 we scaled this 1994 shipment value to decline according to the annual decrease in magnetic ballast shipments projected in the NAECA ballast standards analysis. Beginning in 2011 we made the simplifying assumption that T12 lamp shipments ceased.

Assumptions for unit wattage savings, product service lifetime, operating hours, and market shares by lamp type and by new vs. renovation market are from DOE's 2000 fluorescent lamp ballast standards analysis. Lamp prices are from the 1992 Lighting Policy Analysis by Atkinson *et al.*[10]

Incandescent Reflector Lamps

We estimated the impacts of the incandescent reflector lamp standards from 1996 – 2015. (The standards took effect on November 1, 1995, so we assumed that savings began in 1996.) We used shipments data from past and recent analyses to estimate the annual shipments of lamps complying with the standards. For the commercial sector, complying shipments were derived for 1996 - 2000 from the 1992 Lighting Policy Analysis (Atkinson *et al.*), for 2006 - 2015 from DOE's 2009 incandescent reflector lamp standards NIA spreadsheet (DOE 2009),[15] and for 2001– 2005 by linear interpolation. For the residential sector, we estimated complying shipments for 1995 as 10 percent of total shipments, for 2001 – 2015 from DOE 2009, and for 1996 to 2000 by linear interpolation. Assumptions for unit wattage savings are from Atkinson *et al.* Product service lifetime and operating hours are from DOE 2009. Lamp prices are from Atkinson *et al.*

PLUMBING PRODUCTS

For showerheads and faucets, we started with product lifetime, 2010 site energy savings (from reduced hot water use), and 2010 water savings from Koomey *et al.*[11] For toilets, we used product lifetime and 2010 water savings.

From the energy savings for 2010, we estimated both the site and source energy savings over a 20-year period starting from the compliance date, assuming that shipments retire uniformly over the lifetime and are replaced (constant annual sales).

Based on this uniform retirement function, the energy savings from units surviving beyond the 20-year period are also calculated. We performed this same estimation for the annual time series of water savings.

We derived operating cost savings by applying annual time series of average fuel prices and water prices to the site energy and water savings.

We estimated that there is zero unit incremental cost for these products because, when manufacturers first started to comply with EPACT 1992, they generally did not make significant changes to the products.

The estimates only cover residential use because no data were available to estimate commercial sector impacts.

EPACT 2005

For all of the standards except commercial air conditioners (AC) and heat pumps, we started from the following data reported by Nadel *et al.*[3] for each standard: (1) site energy savings in 2020 and 2030, (2) cumulative energy savings through 2030, (3) NPV for products sold through 2030, (4) lifetime, (5) unit annual energy saving, and (6) unit incremental equipment cost. Nadel *et al.* used a constant efficiency base case, but they also did not model any increase in shipments; these two factors would counteract to some extent.

From the energy savings for 2020 and 2030, we estimated both the site and source energy savings for 25 years of shipments starting from the compliance year.

Using the energy savings per unit and the annual energy savings, we calculated the shipments in each year.

Once we derived the shipments, we could calculate the total incremental equipment cost.

We accounted for impacts to shipments through 2030. The number of years of shipments ranges from 21 to 25, depending on the particular standard.

For commercial AC and heat pumps, DOE National Impact Analysis spreadsheets were available. For these products, we followed the methods described in the DOE Standards 2007- 2010 section.

EISA 2007

For most EISA 2007 standards, we started from the following data for each product reported by DOE in its technical report:[4] (1) cumulative energy savings (through 2038), (2) NPV at 3-percent and 7-percent discount rates. From other relevant DOE sources, we obtained the lifetimes of the products. The DOE report used a constant efficiency base case, which may tend to somewhat overestimate the savings from the standards. To compensate, we used 25 years of shipments instead of 30 years.

We assumed that units retire uniformly over the lifetime and that the annual energy savings will go up after the compliance date until it stabilizes when all the pre-standard units have been replaced by units meeting the standards. The period that it takes for the annual energy savings to reach its maximum is equal to the lifetime of the product. Using these assumptions, we calculated the annual site and source energy savings that will match the given cumulative energy savings. Then we used the Excel Solver to solve for the unit energy savings and incremental equipment cost per unit that will give an NPV that closely matches the given NPV at a 7-percent discount rate. We then adjusted the calculations to account for 25 years of shipments.

For a few EISA 2007 standards (residential boilers, dishwashers, and dehumidifiers), DOE National Impact Analysis spreadsheets were available. For these products, we followed the methods described in the following DOE Standards 2007-2010 section.

DOE STANDARDS 2007-2011

We used the Final Rule NIA spreadsheets from the DOE rulemakings for each of these standards. We set up the spreadsheets for the compliance year and standard levels that were selected in the Final Rules. This gave the annual time series desired: primary energy savings, additional installed cost, and operating cost savings.

In some cases, the time series presented in the spreadsheets were by individual product classes, so we summed them to arrive at totals for the product category or categories in question. In some cases we also made modifications to the spreadsheets to arrive at consistent results across products—for instance, always using 30 years of shipments and extending energy cost savings and energy savings to the end of the lifetime of the units shipped in the 30th year.[16]

General Methods

We converted dollars from the year given in the various sources to 2011$ using the Consumer Price Index for residential appliances and the GDP implicit price deflator for commercial and industrial equipment.

The annual average CO_2 emissions factors for the electricity generation sector are derived for each year through 2008 from EIA statistics on total CO2 emissions from the electric power sector and total primary energy consumption by the electric power sector and, for each year after 2008, are derived from similar forecasted outputs in ETA's *Annual Energy Outlook 2011.* The social cost of carbon (SCC) values from 2010 onward are from the mid-range series developed by a Federal interagency process. This series is based on the average SCC derived from the three integrated assessment models that were examined, using a 3-percent discount rate.[17] Because the SCC values are based on a 3-percent discount rate, we used the same discount rate to discount the future annual estimates of the value of reduced CO_2 emissions. SCC values prior to 2010 were estimated based on the post-2010 trend.

Appendix B: Quad and Carbon Equivalents

From the U.S. Department of Energy's 2010 Buildings Energy Data Book.

1.5.1 Key Definitions

Quad: Quadrillion Btu (10^15 or 1,000,000,000,000,000 Btu)

Generic Quad for the Buildings Sector: One quad of primary energy consumed in the buildings sector (includes the residential and commercial sectors), apportioned between the various primary fuels used in the sector according to their relative consumption in a given year. To obtain this value, electricity is converted into its primary energy forms according to relative fuel contributions (or shares) used to produce electricity in the given year.

Electric Quad (Generic Quad for the Electric Utility Sector): One quad of primary energy consumed at electric utility power plants to supply electricity to end-users, shared among various fuels according to their relative contribution in a given year. (Note: The consumption of an electric quad results in the delivery of just under 1/3 the electric quad due to generation and transmission losses.)

Primary Energy: The total energy consumed by an end-user, including the energy used in the generation and transmission of electricity. Also referred to as "source" energy.

Delivered Energy: The energy consumed by an end-user on site, not including electricity generation and transmission losses.

1.5.2 Consumption Comparisons in 2008

One quad equals:

- 49.5 million short tons of coal
  - = enough coal to fill a train of railroad cars 4,066 miles long (about one and a half times across the U.S.)
- 972.8 billion cubic feet natural gas
- 8.2 billion gallons of gasoline = 21.3 days of U.S. gasoline use
  - = 22. million passenger cars each driven 12,400 miles
  - = 19.2 million light-duty vehicles each driven 12,200 miles
  - = all new passenger cars and light-duty trucks sold, each driven 40,000 miles
  - = 14. million stock passenger cars, each driven 11,500 miles = 10% of all passenger cars, each driven 11,500 miles
  - = all new passenger cars each making 9 round-trips from New York to Los Angeles
- 172.4 million barrels of crude oil = 14.26 days of U.S. imports = 167 days of oil flow in the Alaska pipeline at full capacity
  - = the amount of crude oil transported by 483 supertankers
- 17.8 hours of world energy use
- the electricity delivered from 238 coal-fired power plants (200-MW each) in one year
- the electricity delivered from 37 nuclear power plants (1000-MW each) in one year
- average annual per capita consumption of 3.06 million people in the U.S.
- the approximate annual primary consumption of any one of the following states: Arkansas, Iowa, Kansas, Maryland Massachusetts, Mississippi, South Carolina, Utah, Washington, or Wyoming (1)

Note(s): 1) All states listed have annual energy consumption that is within 20% off one quad.

Source(s): EIA, Annual Energy Outlook 2011 Early Release, Dec. 2010, Summary Reference Case Tables, Table A1, p. 1-2, Table A2, p. 3-5, Table A7, p. 34-35, Table A8, p. 18-19, Table A9, p. 20-21, and Table A11, p. 23-24 for consumption; EIA, Annual Energy Outlook 2010, May 2010, Table G1, p. 215 for heat rates; EIA, State Energy Data 2008: Consumption, June 2010, Table S3, p. 5, Table R1, p. 15, and Table R2, p. 16; EIA, Electric Power Annual 2009, January 2011, Table 1.1, p. 14; DOC, Statistical Abstract of the United States 2008, May 2008, No. 1080 p. 690; DOC, Statistical Abstract of the United States 2011, 2010, No. 1031, p. 658, and No. 1074, p. 686; and Newport News Shipbuilding Web site.

1.5.3 Carbon Emission Comparisons

One million metric tons of carbon dioxide-equivalent emissions equals:

- the combustion of 525 thousand short tons of coal
- the coal input to 1 coal plant (200-MW) in ten and a half months
- the combustion of 18 billion cubic feet of natural gas
- the combustion of 118 million gallons of gasoline = the combustion of gasoline for 7 hours in the U.S.
  - = 310 thousand new cars, each driven 12,400 miles
  - = 272 thousand new light-duty vehicles, each driven 12,200 miles
  - = 260 thousand new light trucks, each driven 11,000 miles
  - = 0.14 million new passenger cars, each making 5 round trips from New York to Los Angeles
- the combustion of 190 million gallons of LPG
- the combustion of 107 million gallons of kerosene
- the combustion of 102 million gallons of distillate fuel
- the combustion of 87 million gallons of residual fuel
- 17 minutes of world energy emissions
- 90 minutes of U.S energy emissions
- 3.9 hours of U.S. buildings energy emissions
- 7 hours of U.S. residential energy emissions
- 8 hours of U.S. commercial energy emissions
- 1 day of U.S. buildings lighting energy emissions
- average annual per capita emissions of 52,000 people in the U.S.

Source(s): EIA, Annual Energy Outlook 2011 Early Release, Dec. 2010, Summary Reference Case Tables, Table A2, p. 3-5, Table A7, p. 16-17 for consumption and Table A18, p. 36 for emissions; EIA, Annual Energy Outlook 2010, May 2010, Table G1, p. 221 for heat rates; EIA, Electric Power Annual 2009, January 2011, Table 1.2, page 17; EIA, Country Energy Profiles for global emissions, available at http://www.eia.gov/country/index.cfm, accessed 2/3/2011; EIA, Assumptions to the Annual Energy Outlook 2010, May 2010, Table 1.2, p. 12 for carbon coefficients; and DOC, Statistical Abstract of the United States 2008, Jan. 2008, No. 2, p. 8 and No. 1084, p. 715.

References

[1] Meyers, S., J. McMahon, and B. Atkinson. Realized and Projected Impacts of U.S. Energy Efficiency Standards for Residential and Commercial Appliances. LBNL-63017. Lawrence Berkeley National Laboratory, Berkeley, CA. 2008. http://www.osti.gov/bridge/purl.cover.jsp?purl=/938510-IKYBEw/

[2] Nadel, S., A. deLaski, J. Kleisch, and T. Kubo. Leading the Way: Continued Opportunities for New State Appliance and Equipment Efficiency Standards. American Council for an Energy-Efficient Economy and Appliance Standards Awareness Project. Washington, DC. January 2005. http://www.aceee.org/sites/default/files/publications/researchreports/a051.pdf

[3] 3 TECHNICAL SUPPORT DOCUMENT-Impacts on the Nation of the Energy Independence and Security Act of 2007, March 2009; Available at: http://www1.eere.energy.gov/buildings/appliance standards/pdfs/en masse tsd march 2009. Pdf

[4] Office of Management and Budget. OMB Circular A-4, Regulatory Analysis. Washington, D.C. September 17, 2003. www.whitehouse.gov/omb/assets/omb/circulars/a004/a-4.pdf

[5] S. Sorrell, J. Dimitropoulos, and M. Sommerville, "Empirical estimates of the direct rebound effect: a review," Energy Policy 37(2009) pp. 1356–71.

[6] Belzer, D. and D. Winiarski. EPACT-Covered Commercial HVAC and Water Heating Equipment: Summary of Energy Savings and Economic Benefits – 2004. 11/4/2004. Pacific Northwest National Laboratory.

[7] Rosenquist, Greg, Michael McNeil, Maithili Iyer, Steve Meyers, and Jim McMahon. Energy Efficiency Standards for Equipment: Additional Opportunities in the Residential and Commercial Sectors. Energy Policy 34(2004) pp 3257-3267.

[8] Nadel, S., H. Geller, F. Davis and D. Goldstein. 1989. Lamp Efficiency Standards for Massachusetts: Analysis and Recommendations. June, 1989. American Council for an Energy-Efficient Economy (Washington DC), Fred Davis Corporation (Medfield MA), Natural Resources Defense Council (San Francisco CA). Prepared for Massachusetts Executive Office of Energy Resources. Research Report A891.

[9] Geller, H., and S. Nadel. Consensus National Efficiency Standards for Lamps, Motors, Showerheads and Faucets, and Commercial HVAC Equipment. American Council for an Energy-Efficient Economy. Washington, DC. June 1992. http://www.aceee.org/sites/default/files/publications/researchreports/A921.pdf

[10] B. Atkinson et al., Analysis of Federal Policy Options for Improving U.S. Lighting Energy Efficiency: Commercial and Residential Buildings. LBL-31469. December 1992. Lawrence Berkeley Laboratory.

[11] Koomey, Jonathan, Camilla Dunham, and James D. Lutz. The Effect of Efficiency Standards on Water Use and Water Heating Energy Use in

the U.S.: A Detailed End-use Treatment. LBL35475. Lawrence Berkeley Laboratory. May 1994.

End Notes

[1] The term "consumer" as used in this report refers to all buyers and users of appliances and equipment covered by standards.

[2] Dishwashers, residential boilers and dehumidifiers.

[3] For details on the SCC values, see appendix 16-A of the Technical Support Document for the final rule for residential refrigeration products. http://www1.eere.energy.gov/buildings/appliancestandards/pdfs/refrig finalrule tsd.pdf

[4] The increase in total energy use depicted after 2025 is due to growth in purchases of refrigerator-freezers. If the standard is updated as required by EPCA, the declining trend would continue.

[5] http://www1.eere.energy.gov/buildings/appliance_standards/pdfs/refrig_ fi nalrule_tsd.pdf

[6] Additional results, including impacts by each standard, are presented in the Excel spreadsheet used to calculate the impacts, which may be found at the page for this report at: http://efficiency.lbl.gov/bibliography/

[7] Note that water savings estimates are not available for commercial plumbing products (water closets, urinals, and faucets).

[8] USGS estimates that water withdrawals for public supply were 44.2 billion gallons/day in 2005. http://pubs.usgs.gov/circ/1344/

[9] The calculation uses an energy use value of 3,000 kWh per million gallons, which is an estimated average for the U.S. based on several sources.

[10] Consumers also saved energy from lighting products standards, but we were not able to disaggregate the estimated savings by these standards by end-use sector.

[11] In 2011 DOE began to account for change in product prices in its forecasts.

[12] See for example: http://www.dalemarsden.ca/docs/publications/Marsden_ etal_2006.pdf

[13] DOE does not adjust the energy cost savings for the rebound effect because it believes that, if it were able to monetize the increased value to consumers associated with the rebound effect, this value would be similar to the foregone energy savings.

[14] Further details are given in cell comments in the Excel spreadsheet used to calculate the impacts, which may be found at the page for this report at: http://efficiency. lbl.gov/bibliography/

[15] See: http://www1.eere.energy.gov/buildings/appliance_standards/ resi dential/incandescent_lamps_standards_final_rule_to ols.html

[16] This accounting approach differs from what DOE reports in its rulemakings, in which DOE presents energy savings over a 30-year period.

[17] For more information on these SCC values, see chapter 16 and appendix 16-A of the Residential Clothes Dryers and Room Air Conditioners Direct Final Rule Technical Support Document. http://www1.eere.energy.gov/ buildings/appliance_standards/residential/ resi dential_clothes_dryers_room_ac_dir ect_final_rule_tsd.html

In: Climate, Energy, and Environment. Vol. 1 ISBN: 978-1-62948-407-5
Editor: Jake S. Diaz

Chapter 5

KLAMATH BASIN SETTLEMENT AGREEMENTS: ISSUES IN BRIEF*

Charles V. Stern, Harold F. Upton, Cynthia Brougher and Betsy A. Cody

SUMMARY

The Klamath River Basin on the California-Oregon border is a focal point for local and national discussions on water allocation and species protection. Previously, water and species management issues have exacerbated competition and generated conflict among several interests—farmers, Indian tribes, commercial and sport fishermen, federal wildlife refuge managers, environmental groups, and state, local, and tribal governments. Drought conditions and a recent call for water by senior water rights holders have again brought these issues to the forefront.

In 2010, the Secretary of the Interior and the governors of Oregon and California, along with multiple interest groups, announced the result of multi-year negotiations in an effort to resolve long-standing issues in the basin: two interrelated agreements, supported by the federal government and signed by the two states and numerous other parties. These agreements, known as the Klamath Basin Restoration Agreement (KBRA) and the Klamath Hydroelectric Settlement Agreement (KHSA),

* This is an edited, reformatted and augmented version of a Congressional Research Service publication, CRS Report for Congress R42158, from www.crs.gov, prepared for Members and Committees of Congress, dated June 18, 2013.

together aim to provide for water deliveries to irrigators and wildlife refuges, fish habitat restoration, and numerous other related actions. Generally, the KBRA provides for actions intended to restore Klamath fisheries and for assurances for water deliveries to wildlife refuges and federal project irrigators under certain circumstances, among many other things. The KHSA lays out a process that could potentially lead to the removal of four privately owned dams on the Klamath River. This dam removal would be one of the largest and most complex projects of its kind ever undertaken.

Some parts of the Klamath Agreements are already being carried out under existing authorities. Studies to inform a Secretarial Determination on dam removal under the KHSA have been completed, and some restoration actions have already begun. However, new authorization from Congress is required in order for the most significant components of the agreements to be implemented. Other ongoing processes in the Klamath Basin, such as the recent water rights adjudication by the State of Oregon, may result in significant changes in water allocations in the basin with or without federal legislation to authorize the Klamath Agreements. However, these developments are related.

The Klamath Agreements were originally set to expire in 2012 if no authorizing legislation was enacted, but have been extended to 2014. Legislation in the 112th Congress would have authorized the agreements, including hundreds of millions of dollars in federal actions under the KBRA and authorization for the Secretary of the Interior to make a "final" dam removal recommendation under the KHSA. No such legislation has been proposed in the 113th Congress.

Congressional consideration of the Klamath Agreements could include whether the federal government should act beyond its current activities in the Klamath Basin, the type of strategies proposed in the agreements and their cost, and the potential for the approach in the Klamath Basin to serve as precedent for the resolution of similar conflicts in other parts of the country.

INTRODUCTION

The Klamath River Basin, a region along the California-Oregon border, has become a focal point for local and national discussions on water resources and species management. Water management issues were brought to the forefront when severe drought conditions in 2001 exacerbated competition for scarce water resources and generated conflict among several interests—farmers, fishermen (commercial and sport), other recreationists, federal wildlife refuge managers, environmental organizations, and state, local, and tribal governments. Subsequent problems with Klamath Basin fisheries, in particular events in 2002 and 2006, exacerbated these conflicts. Recent low water conditions and a call for water by senior water rights holders have again brought these issues to the forefront.

Congress has oversight authority over federal activities in the Klamath Basin related to operation of the Bureau of Reclamation's Klamath Project, management of federal lands (including six national wildlife refuges, managed by the Fish and Wildlife Service); and implementation of Endangered Species Act (ESA) and other federal laws.[1] Previously Congress has held hearings and appropriated funding to address issues in the Klamath Basin. Past congressional debate has generally focused on the role of the ESA in water management, the operation of the Klamath Project, and other topics such as supplemental support for parties impacted by federal policies.

The Klamath Basin Restoration Agreement (KBRA) and the Klamath Hydroelectric Settlement (KHSA), collectively referred to as the "Klamath Agreements" in this report, were signed in 2010 by a wide array of basin interests (although not all basin interests support the agreements). They aim to address many of the ongoing conflicts in the Klamath. The KBRA would, among other things, set limits for water allocations for irrigators and wildlife refuges under a range of conditions related to the amount of water forecast in a given year; attempt to make available additional water and power supplies in the basin; and provide for restoration and monitoring of certain fish species. It also includes other assurances to settle ongoing water conflicts between basin tribes and other entities.

The KHSA lays out a process that could lead to removal of four nonfederal hydroelectric dams currently owned and operated by a private entity, PacifiCorp. If carried out as envisioned, the project would be one of the largest, most complex dam removals in history. Under the KHSA, the Secretary of the Interior led a study process to determine whether removal of these dams is in the public interest. Most of this study process has been completed, but the final step in the process, a determination by the Secretary, requires authorization by Congress.

In addition to dam removal, the Klamath Agreements require congressional authorization to move forward on other key components. Consideration of the agreements could result in Congress revisiting previous issues it has discussed related to the Klamath, as well as new ones. This report focuses on congressional consideration of the Klamath Agreements.

Source: Bureau of Reclamation, adapted by CRS.

Figure 1. Klamath River Basin.

BACKGROUND

The Klamath River Basin is a sparsely populated area on the Oregon and California border with limited water resources (see *Figure 1*). Irrigated agriculture in the upper basin relies largely on water provided by the Bureau of Reclamation's Klamath Project. Other farmers and ranchers ("non-project" irrigators) also rely on basin water supplies for irrigation. The area is home to six national wildlife refuges that rely on the same water supplies to sustain migratory bird populations, and several Native American tribes who were historically dependent on lower and upper basin fish species. Two species of upper basin fish are currently listed as endangered under the Endangered Species Act (the Lost River and short nose sucker), and one species of lower basin fish is listed as threatened under ESA (coho salmon, an anadromous fish).

The basin also includes seven dams on the Klamath River and its tributaries, built between 1918 and 1962. Six of these dams are owned by PacifiCorp, a private company, and are known collectively as the Klamath Hydroelectric Project.[2] The dams produce hydroelectric power for the basin, including power for irrigators and other areas.[3] The original FERC license to operate the KHP expired in 2006, and PacifiCorp applied for relicensing of the KHP in 2004. To date, a new long-term license has not been granted for the project because of the lack of state certification under Section 401 of the Clean Water Act, as well as ongoing uncertainty related to fish passage upgrades and agreements discussed below. PacifiCorp is operating the dams on a temporary annual license until these issues are solved.

PREVIOUS EVENTS

While water and species management issues have been prevalent throughout the history of the Klamath Project, recent congressional consideration related to Klamath Basin issues usually centers on one or more of three seminal events in 2001, 2002, and 2006. These events resulted in press coverage, legal conflicts, and studies which framed the negotiations and agreements that are currently at issue. Most recently, in the summer of 2013 a call on water rights resulted in limitations on water deliveries for junior water rights holders in the upper part of the basin.

Several events put the Klamath region in the national spotlight. In 2001, as a result of previous biological opinions by the federal Fish and Wildlife Service (FWS) and the National Marine Fisheries Service (NMFS), Reclamation severely curtailed water deliveries to the Klamath Project to provide more water for endangered fish in the basin and prevent their extinction.[4] These allocations were met with protests by area irrigators, who threatened to open irrigation head gates by force.[5] Later, in 2002, irrigators received more water than had been allocated in 2001, but thousands of fish (mainly Chinook salmon) died on the lower part of the Klamath River, largely due to poor water conditions and fish health in that part of the basin. In 2006 NOAA severely restricted ocean fishing for salmon in the region due to low numbers of naturally spawning adults (due in part to residual effects of the aforementioned 2002 event). This resulted in a large decrease in that year's commercial and recreational salmon catch compared to previous years.

The federal government provided emergency funding in response to these and other events in the Klamath. The funding included at least $170 million in addition to regular programmatic expenditures over the last decade. For instance, for the 2001 and 2006 events, the federal government provided approximately $35 million and $60 million in emergency aid, respectively.[6]

Aid in addition to regular agency programs and appropriations was also provided in other years. Between 2002 and 2007, Reclamation spent $14 million on a pilot water bank for the Klamath to alleviate water shortages.[7] Due to drought events in 2010, an additional $10 million in supplemental appropriations was provided to the Klamath Basin in that year, and $2 million was provided for a Klamath Drought Initiative by the U.S. Department of Agriculture (USDA).[8] The 2002 farm bill provided $50 million to the Klamath Basin, and USDA funding was also provided under other general authorities and programs authorized in the 2002 and 2008 farm bills.[9]

Klamath Water Rights Adjudication

The questions related to the quantification of tribal water rights are interconnected with the determination of water rights within the Klamath Basin. The Klamath Basin is "over-allocated," meaning that claims to water exceed the amount available in most years. This often leads to legal conflicts over the proper allocation of limited resources. Allocation of water resources is largely determined by state law.

Western states generally follow a system of prior appropriation, which provides certain quantities of water to water users depending on their relative seniority in acquiring water rights. State appropriative rights can be complicated by federal water rights such as those of tribes claiming water rights reserved by the creation of a tribal reservation. In addition to tribal reserved water rights, other federal rights, such as those associated with federal land reservations like national forests and national wildlife refuges, also may not be quantified. The uncertainties resulting from the lack of quantification of these rights has led to ongoing legal disputes over the allocation of water within the Klamath Basin.

Oregon undertook a general adjudication of water rights in the Klamath Basin (known as the Klamath Basin Adjudication (KBA)) to address these disputes. The KBA began in the 1970s to determine water rights among various users in the Klamath Basin.

The general process of the adjudication is as follows: parties with claims or contests must file with the Oregon Water Resources Department (OWRD); an administrative panel then hears the contests and issues proposed orders based on the hearing; and the OWRD reviews the proposed orders and issues its final findings and order, which is filed with a state court. The OWRD's final findings and order were filed with the state court in March 2013. The order generally upheld previous claims and determined that the most senior claims in the basin are held by the United States in trust for the Klamath Tribes, and carry a priority date of "time immemorial." Even with the conclusion of the administrative adjudication, parties that are dissatisfied with the outcome may pursue judicial appeals. For example, following the OWRD's final determinations, parties may file "exceptions to the Determination" with the state court. Following the state trial court's decision, litigants may appeal through the state's court of appeals, state supreme court, and possibly the U.S. Supreme Court. Thus, although the KBA has been completed and announced, many observers expect that appeals may be made and that disputes over the allocation of the Klamath Basin's water resources will continue. This may be especially true now that senior water rights holders have made a "call" on water, which may affect junior water rights holders.

Sources: Or. Rev. Stat. 539.010 et seq., United States v. Braren, 338 F.3d at 973-74. All documents related to the adjudication are available at http://www.oregon.gov/owrd/Pages/adj/index.aspx.

In addition to the aforementioned events, an ongoing state water rights adjudication in Oregon, which was first initiated in the 1970s, received added attention in 2013.

The State of Oregon upheld a number of previous water rights claims, including, among other things, that the Klamath Tribes have water rights with a date of "time immemorial." Significantly, these water rights may now be exercised over junior water rights holders in the basin to limit their water deliveries. The ongoing adjudication process is a major factor in Klamath settlement negotiations. (See box above, "Klamath Water Rights Adjudication.")

KLAMATH SETTLEMENT AGREEMENTS

In response to conflicts and other issues in the Klamath basin, the federal government facilitated talks among multiple groups between 2002 and 2010, including formal negotiations to reach "settlement" agreements between 2006 and 2010.[10] Participants in negotiations included state governments, tribes, counties, irrigators, fishermen, and conservation groups. The goal of the negotiations was a long-term solution to the multiple water and endangered species issues in the Klamath Basin, including the aforementioned issues associated with irrigation deliveries and flows for fish, as well as potential issues associated with pending water rights adjudications and relicensing of the basin's hydroelectric dams. The two agreements that resulted from the negotiations, the KBRA and the KHSA, are officially linked in that signatories officially see them as complementary and do not support authorization of one agreement without the other.[11] Originally, both agreements were set to expire without congressional authorization by 2012, but this deadline has since been extended to 2014.

Klamath Basin Restoration Agreement

The KBRA was negotiated by stakeholders and other groups in the Klamath Basin. It was agreed to by more than 40 signatories, or "parties," although not all interests in the basin support it. Broadly speaking, under certain conditions, parties to the KBRA promise to support diversions for Klamath Project irrigators and federal wildlife refuges that correlate to a given year's forecast inflows into Upper Klamath Lake. Water that is "surplus" to these inflows and not subject to other valid water rights (e.g., off-project diversions) would be allocated to other uses, including instream flows (see *Table 1* for a summary of this arrangement). In exchange for this support, environmental interests would gain additional federal and state funding for fisheries restoration, some of the aforementioned surplus water supplies, and related assurances for dam removal under the KHSA (which is expected to restore fisheries). Parties have also agreed that implementation of the ESA will not be affected under the agreement. For their part, three of the largest four tribes in the basin agreed to support the project and refuge diversions and not make a call on certain water rights in exchange for the aforementioned restoration actions, federal actions to restore fisheries, and economic aid.

Table 1. Water Allocations in the Klamath Basin Restoration Agreement

		Water Forecast[a]	
	"Dry" Year[b]	"Average" Year[c]	"Wet" Year[d]
Klamath Reclamation Project	March-Oct.: 330,000	March-Oct. (formula-based): 330,000-385,000	March-Oct.: 385,000
	Nov.-Feb.: 45,000		Nov.-Feb.: 45,000
		Nov.-Feb.: 45,000	
Wildlife Refuges	March-Oct.: 48,000	March-Oct. (formula-based): 48,000-60,000	March-Oct.: 60,000
	Nov.-Feb.: 35,000		Nov.-Feb.: 35,000
		Nov.-Feb.: 35,000	
Environmental/Other[e]	n/a	n/a	n/a

Source: Klamath Basin Restoration Agreement, Appendix E-1, p. E-25.

Notes: Units in whole acre-feet (a/f). Columns indicate the water allocation under a given forecast scenario. Rows indicate the diversion reserved for a specific location.

[a] Forecast references the March 1 Natural Resources Conservation Service Forecast for Net Inflow into Upper Klamath Lake for the period April 1-September 30.

[b] "Dry" indicates inflows less than 287,000 a/f. Section 19.2.2.B.v of the KBRA provides that if an "Extreme Drought" is declared by OWRD and voluntary water conservation measures triggered under the KBRA are insufficient, diversions may be reduced below the levels specified in the KBRA.

[c] "Medium" indicates forecast inflows ranging from 287,000 a/f- 569,000 a/f.

[d] "High" indicates forecast inflows of more than 569,000 a/f.

[e] Additional allocations (including environmental flows for fish) are assumed in the KBRA, but not provided with a specific diversion limit/guarantee.

Although exact projections of the effects of the KBRA allocations on various water users are subject to debate, compared to recent water years they are generally expected to result in less water for Klamath Project irrigators in wet years, but more certainty, and potentially greater allocations than may have been the case otherwise, in dry years. Under the agreement, area irrigators are also promised funding to develop low-cost power to replace hydropower previously provided by the PacifiCorp dams, as well as funding to potentially make available more water supplies through means yet to be determined.[12] Irrigators who did not agree to the KBRA, including some off-project irrigators in the upper basin, are not subject to the KBRA's assurances as they relate to tribal water rights. The federal government is not party to the KBRA until Congress enacts authorizing legislation. Many of the actions envisioned by the KBRA have been interpreted as not being authorized, and would require significant federal appropriations to go forward. According to previous estimates by parties to the KBRA, federal costs to implement the Klamath agreements would be $798.5 million or $536 million over 15 years, depending on the assumptions used.[13]

Klamath Hydroelectric Settlement Agreement

The KHSA initiates a process that could lead to removal of four nonfederal hydroelectric dams (J.C. Boyle, Iron Gate, Copco 1, and Copco 2) currently owned and operated by a private entity, PacifiCorp, and provides related assurances.

Most significantly, the KHSA lays out a process for additional studies and environmental review by the Secretary of the Interior to consider removal of the dams (known as the "Secretarial Determination"). Under the agreement, facilities' removal would be paid for by ratepayers in California and Oregon ($200 million) and an assumed California Water Bond ($250 million). The entity responsible for removal (known as the Dam Removal Entity, DRE) has yet to be defined.[14]

The KHSA also addresses the interim operation of the dams as well as proceedings that could lead to transfer, decommissioning, and removal of the dams. The KHSA would transfer one dam (Keno Dam) to the Bureau of Reclamation, and would initiate a process to decommission other resources associated with the project.

In contrast to the KBRA, the federal government was party to the KHSA, and some actions within the KHSA have been interpreted by Reclamation to not require an explicit authorization by Congress (i.e., they are authorized under existing, more general, authorities) and have been completed. The central component of the KHSA, the dam removal study, was conducted under Reclamation's general authorities and finalized in October 2012.[15] DOI and most parties agree that congressional authorization is necessary to make a final Secretarial Determination on dam removal and to move forward with that project.[16]

The Secretarial Determination was originally expected by March 2012, but since no authorizing legislation has been enacted, the deadline (which was not binding) has passed.

CONGRESSIONAL INTEREST

Both Klamath Agreements require congressional authorization to move forward. Companion bills to authorize the Klamath Agreements were introduced in the 112th Congress in the House and Senate (H.R. 3398 and S. 1851). The bills would have authorized most of the agreements by reference, and explicitly reinforced other provisions in the agreements. No such legislation has been introduced in the 113th Congress.

In considering legislation related to Klamath restoration, Congress may focus on a number of issues, including the cost of the legislation and the advisability of the array of specific strategies and actions in the two agreements (e.g., dam removal, payments to irrigators to forgo diversions, etc.) versus the status quo and the potential for future disruptions in water supply to irrigators, to fish and wildlife, and to tribes in the basin.

Role of the Federal Government

The role of the federal government in the Klamath Basin centers largely on operation of the Bureau of Reclamation's Klamath project, management of several national wildlife refuges and other fish and wildlife resources under the ESA, and tribal trust responsibilities.[17] The federal role has been contentious in the past and is a central question related to congressional consideration of the Klamath Agreements. Both agreements assume numerous actions by the federal government, and as noted earlier, many of these actions cannot go forward absent new congressional authorization. In particular, without congressional authorization for the Secretary to make a dam removal determination, the four dams could not be removed under the process currently envisioned in the KHSA. Further, without authorization of its provisions, many of the programs under the KBRA could not go forward, such as fisheries restoration, water rights retirement, and other actions which allowed for the widespread support for the KBRA's water allocations.

Supporters of the Klamath Agreements argue that because of the federal government's prominent role in the basin, including its role in the area's resource allocation conflicts, it has a responsibility to help solve these issues. These groups note that federal involvement, including operation of the Klamath Project, implementation of ESA, and management of fisheries and federal lands, is central to the issues in the basin. They argue the agreements represent a consensus achieved by majority of basin interests that are traditionally opposed to one another, and are the best opportunity to solve the region's problems going forward. They also argue that many of the expenses for these agreements are likely to be offset in the form of reduced future federal expenditures for litigation and emergency financial support. Supporters also note that the agreements will be a valuable source of jobs within the basin.[18]

Opponents of one or both of the agreements cite a number of reasons for their opposition. Some of those opposed to the agreements note that the federal government has no clear obligation to authorize and implement the agreements. Many of these entities have argued that the activities represented in both agreements, including dam removal and water quality improvements, could potentially occur without the federal involvement through pre-existing processes.[19] Others, including off-project water users and some local interests, note that the agreements, like previous federal actions in the Klamath, amount to federal overreach and are likely to harm the local economy, especially the agricultural and recreation industries. Finally, others argue against federal authorization of the settlement agreements because they believe that specific components of the agreements will undermine existing federal laws (e.g., the Endangered Species Act) and/or federal responsibilities (e.g., tribal trust responsibilities), or that they will fail to achieve their stated goals (e.g., fisheries restoration).

Cost of Implementation

The cost to the federal government to implement actions proposed under the Klamath agreements is contentious. Most of the actions in the KBRA are funded by the federal government, including most of the costs for the water resources, fisheries restoration, and tribal components of the agreement. The original 2010 KBRA included an estimate of $970 million in total costs from 2010 to 2020.[20] Since that time, estimates have been revised downward to $798.5 million or $536 million, depending on which assumptions are used.[21]

Cost estimates to implement the KHSA have not changed substantially since the original agreement, and may not garner as much attention from Congress since states are the primary entities responsible for funding dam removal under the KHSA.[22] Previous studies by DOI to inform the Secretarial Determination estimated potential costs of approximately $290 million for dam removal, which is less than the amount anticipated to be available from ratepayers ($200 million) and the State of California ($250 million).[23] One potential issue related to these costs is whether the Secretary will recommend the federal government as the dam removal entity. If so, a major question may be how the department would handle additional costs for dam removal, such as costs resulting from lawsuits.

There is no formal estimate of potential future savings to state and federal governments associated with the agreements. Supporters point to previous costs to federal and state governments, including at least $170 million in additional aid for irrigators and fisheries beyond "regular" appropriations provided from 2001 to 2011.[24] Additionally, advocates point to decreased costs for litigation. Although the agreements would not prevent future litigation, supporters argue that if authorized, they would obligate parties to pursue other dispute resolution mechanisms and would thus render future litigation costs less likely. Opponents note that none of these savings are guaranteed under the agreements, and that supplemental appropriations and expenditures for litigation may still be necessary.

Supporters also argue that in addition to potential savings, the KBRA and KHSA could create significant economic benefits (in terms of both traditional and "non-use" benefits). Studies commissioned as part of the dam removal process by DOI estimated the potential value of restoration, including non-use values.[25] The department estimated the net value of restoring the Klamath at $16 billion-$84 billion, depending on the assumptions and methodology used.[26] Some dispute these estimates, seeing them as unrealistically large and derived from questionable methodologies.

Obtaining authorization and appropriations for these activities from Congress may be difficult. For new authorizations, there are procedural hurdles that need to be overcome in the House. New authorizations may not be eligible for consideration without an "offset" of another authorization, pursuant to House "Cut-Go" protocol. Beyond authorization of new funds, some observers also note that it is unlikely that DOI will obtain the new appropriations envisioned for the KBRA in a constrained budgetary environment. Hypothetically, a lack of discretionary appropriations or overall progress associated with future actions initially assumed in the KBRA could affect the status of support for either agreement among the parties, and thus cause additional conflicts among the agreement's supporters. Supporters have for the most part acknowledged these difficulties, but argue that initial authorization of the agreements is an important first step, and that the basin's issues are important enough to warrant congressional authorization and funding.

Dam Removal

Some believe that congressional authorization of the agreements would be an implicit endorsement of dam removal. Since authority for the Secretary of the Interior to make a final determination on dam removal is a key step in the KHSA, some argue that the authorization process is the primary opportunity for Congress to weigh in for or against this decision. The report findings that were previously sent to the Secretary did not find major drawbacks associated with dam removal, and many have concluded that a positive finding by the Secretary would be likely if congressional authorization were provided. While many in Congress support dam removal as a means to restore rivers, others see it as an unnecessary and expensive step that decreases the availability of renewable energy.

Additionally, while the KHSA anticipates that ratepayers and the State of California will be the primary entities funding dam removal, it remains unclear whether the federal government will be involved in this process and, if so, how.

There is considerable disagreement among stakeholders on whether dam removal would happen without the KHSA and the related determination by the Secretary of the Interior. Opponents have noted that dam removal, funded by PacifiCorp, would be likely under FERC relicensing. However PacifiCorp and supporters of the KHSA note that a different outcome (such as fish passage upgrades) could occur under the FERC process for a number of reasons (for more information, see below section, "Stakeholder Views").

Science and the Klamath Agreements

The science underpinning water allocation and other decisions in the Klamath Basin has been contentious. Most prominently, the biological opinions by the Fish and Wildlife Service (FWS) and the National Marine Fisheries Service (NMFS), which led to the 2001 decision to not make deliveries available to the Klamath Project, were extremely controversial. Reclamation's 2002 decision to reject the NMFS and FWS opinions was similarly controversial. These decisions were the subject of a 2004 National Research Council review which concluded that scientific data were insufficient to support the FWS and NMFS management regimes that had been proposed for Upper Klamath Lake for the 2001 growing season. However the review found support for other measures in the biological opinions.[27]

The science underpinning the KBRA and the KHSA has also been criticized. The KHSA underwent lengthy peer review and public comment processes, but was the subject of a scientific integrity complaint by a Reclamation science advisor who alleged that DOI violated a 2009 executive order by misrepresenting the effect of dam removal on salmon in study summaries, among other things.[28] On January 7, 2013, seven Reclamation biologists filed a separate scientific integrity complaint, alleging that Reclamation officials violated DOI scientific integrity policies by threatening to reassign or eliminate scientists within the Fisheries Resources Branch of the Klamath Basin Area Office.[29] These officials argue that they were targeted for retribution because their science contradicted that of other agencies, such as FWS and NMFS. Finally, a complaint related to the scientific analysis underpinning dam removal was also filed with DOI by the Siskiyou County Board of Supervisors in April 2013.[30]

Implementation of ESA

The extent to which the Klamath Agreements will alter implementation of the Endangered Species Act (ESA) and other federal laws is a matter of disagreement. Previous biological opinions established minimum flows on the Klamath River for coho salmon, as well as actions intended to aid the recovery of Lost River and shortnose suckers. According to supporters, the Klamath Agreements will be considered to the maximum extent practicable under the Endangered Species Act.[31] At the same time, both agreements state that implementation of the agreements shall not affect implementation of the Endangered Species Act by the Department of the Interior or the National Marine Fisheries Service.[32] A perceived conflict between these two objectives, and a lack of clarity on how exactly they will be interpreted and implemented, has resulted in ongoing disagreements among some stakeholders.

While the agreements do not "waive" application of the ESA, some groups that were not signatories argue that certain provisions, in particular the defined water allocations for irrigators, would have the effect of undermining the ESA. These groups note that the allocations for irrigators would provide more water than irrigators received under ESA stipulations during recent "low" water years, and will thus decrease flows from amounts provided under recent biological opinions and provide less water for fisheries. They note that while other processes under the ESA will technically go forward, the assurances in the agreements, if adopted in legislation, could result in additional pressure on regulatory agencies to adopt biological opinions that allow the flows set forth in the proposed Water Resources Program.[33]

Supporters note that the Klamath Agreements would provide for more resources and actions to improve habitat for fish species, which they argue are just as important as the difference in flows that could occur under some scenarios. Improvements under the KBRA, including new fish habitat and improved water quality, are assumed to result in greater fish abundance, which would in turn allow managers to forgo the previous restrictive flows that were provided under the ESA. Furthermore, some believe that the KBRA could in the long term encourage more cooperative actions, which could improve the likelihood of the recovery of listed species and are preferable to previous "top-down" regulatory actions.

Stakeholder Views

Stakeholder views on the Klamath Agreements can broadly be divided into those supporting the agreements and those opposed to one or both of the agreements. However, such a simple characterization may not do justice to the motivation and preferences of many groups. While a majority of interest groups involved in initial settlement negotiations endorsed both agreements, reasons for support among these groups are varied, and in some cases are likely to be contingent on specific parts of one agreement or another (e.g., guaranteed water, dam removal) going forward. Among those opposed to the agreements, reasons for opposition also vary widely, and include reasons ranging from perceived economic damages resulting from the agreements to their overall lack of environmental protections (or effect on implementation of existing laws).

Support for Agreements

Among those supporting the Klamath agreements are all of the parties listed as "non-federal parties" within both the KBRA and the KHSA. For the KBRA, this includes five state agencies in Oregon and California, three tribes, two counties (in both states), 25 parties related to the Reclamation Project and some off-project interests, and several other groups (including environmental interests). These same groups are also party to the KHSA.[34] Other groups and individuals were not "party" to the agreements but have stated their support for them. Notably, supporters who were party to one agreement have agreed to support authorizing legislation for the other (e.g., KBRA signatories have backed enactment of the KHSA), and have generally argued that the agreements themselves must be linked.

The States of California and Oregon, as well as the Obama Administration, support the agreements because, as a whole, they represent a potential solution to the protracted resource conflicts in the Upper and Lower Basins. Government representatives have also pointed to the costs that resulted from previous conflicts in the basin, including supplemental aid, crop insurance, mitigation actions, and litigation costs.

Other groups have chosen to support the agreements not just for their potential to end conflicts in the basin, but also because they include specific provisions that are important to them.[35] For instance, environmental groups have pledged to support the allocations for irrigation absent a similar allocation for fish in exchange for assurances of dam removal under the KHSA and other promised fisheries restoration actions under the KBRA. Among irrigators, those on the Klamath Project have pledged to support restoration provisions and less water in wet years in exchange for benefits from water supplies in dry years that would presumably be higher than under the status quo. Approximately half of the off-project irrigators in the upper basin support the agreements, in some cases because the agreements offer potential alternatives that are preferable to losing water deliveries outright due to status of their junior water rights.[36]

For its part, PacifiCorp supports removal of its four dams under the KHSA because retirement of the dams under the terms of the KHSA reportedly represents a more cost-effective option for its ratepayers than FERC relicensing.[37] Previously there have been disagreements over which option the company would pursue in absence of the KHSA: FERC relicensing for ongoing operations on all four dams (which would entail costly improvements for fish passage, and altered operations for water quality) or a surrender of its license and related decommissioning of some or all of the Klamath hydropower projects.[38] Both options would likely be costly for PacifiCorp and its ratepayers, and could cost the company more than the proposed arrangement under the KHSA. As proposed, dam removal under the KHSA would be funded partially by ratepayers, with the other portion funded by the State of California. The KHSA would allow the company to operate the dams under the current management regime through 2019, through annual extensions to the project's FERC license.

Opposition to Agreements

A number of groups and individuals have opposed the Klamath agreements and now argue against their authorization. Some of these parties were initially involved in settlement negotiations but dropped out for various reasons, while others were not invited to participate in negotiations because they were not seen as representing significant interests. Notable opponents of one or both of the Klamath agreements include Siskiyou County in California, the Klamath Off-Project Water Users Association, the Hoopa Valley and Resighini Rancheria and Quartz Valley tribes, the Northcoast Environmental Center (NEC), Waterwatch of Oregon, Oregon Wild, and others.

Some groups oppose the agreements because they believe they will further damage the region's economy. Some off-project users are opposed to the Klamath agreements and have argued in previous testimony before Congress that the agreement will put farmers (in particular, off-project irrigators) out of business.[39] Siskiyou County has opposed the agreements for a number of reasons, including the assertion by some that the PacifiCorp dams provide flood protection and economic benefits for downstream areas.[40] Some residents and officials in these areas also oppose dam removal because of an expected loss of property taxes associated with certain lands that will lose lake frontage when the dams are removed.[41]

Others argue that the agreements do too little to benefit fisheries, and give up too much to farmers and other interests. For instance, the Hoopa Valley Tribe has been critical of the agreements because they do not provide defined amounts of water for fish, and believe that the federal appropriations for restoration actions are unlikely to be funded. Some have also highlighted the uncertain nature of fisheries restoration in the Klamath under the KBRA. Some believe that these uncertainties were highlighted in expert panels as a part of the larger DOI dam removal study process, but were not adequately acknowledged by DOI in its final studies.[42] Some of these groups favor dam removal, but argue that it could be achieved through the existing FERC relicensing processes and does not need to be tied to the KBRA. They note that by providing annual renewals of the FERC license for the Klamath Hydroelectric Project, the KHSA allows the company to avoid project upgrades (or removal) that would otherwise be paid for by the company and benefit fisheries in the short term.[43] Combined with a lack of performance metrics for fisheries restoration provisions and promises to not make a call on project water rights holders under the KBRA, these opponents assert that the agreements disproportionately benefit PacifiCorp and irrigators at the expense of fisheries.

Some environmental groups oppose other provisions of the Klamath agreements and dropped out of negotiations as a result. Waterwatch of Oregon and Oregon Wild find fault with a number of the provisions in the agreements, including the lack of defined water supplies for fish and the inclusion of lease-land farming on wildlife refuges.[44] Similar to the Hoopa Valley Tribe, these groups have called for voiding the KHSA and resuming water quality certification processes under the Clean Water Act in order to force dam upgrades or removal through a separate process, which they argue will be more expedient and cost less for state and federal taxpayers (i.e., a process to be funded by PacifiCorp and potentially its ratepayers).[45]

End Notes

[1] Two freshwater species (Lost River and shortnose suckers) and one anadromous species (Coho salmon) are listed as endangered under the federal Endangered Species Act (ESA).

[2] Link River Dam, a non-hydroelectric dam, is owned by Reclamation and operated by PacifiCorp.

[3] Power costs for pumping are significant for area irrigators, who rely on the KHP for power. (Unlike other Reclamation projects, there is no power component to the Klamath Project.)

[4] Although Klamath Project irrigators have continued to face uncertainty since then, there has not been a curtailment of water deliveries.

[5] Due to changed water conditions and other supplemental measures, irrigators received partial water deliveries later that year.

[6] Figures for 2001 are based on 2002 estimates by Oregon State University. See William S. Braunworth, Jr., Teresa Welch, and Ron Hathaway, et al., Water Allocation in the Klamath Reclamation Project, 2001, Oregon State University Agricultural Extension Service, Special Report 1037, 2002, p. 267, http://extension.oregonstate.edu/catalog/html/sr/ sr1037-e/. This includes approximately $20 million in aid that was provided from USDA under the Supplemental Appropriations Act, 2001 (P.L. 107-20), $2.2 million provided from BOR for payments to farmers for groundwater, and an additional $13 million in USDA funding provided under other emergency authorities, including crop insurance. For the 2006 fishery disaster declaration, the full funding amount was provided to through the Commerce Department under the U.S. Troop Readiness, Veterans' Care, Katrina Recovery, and Iraq Accountability Appropriations Act, 2007 (P.L. 110-28).

[7] Personal correspondence, Bureau of Reclamation, June 6, 2012.

[8] The 2010 funding was provided to Reclamation under the Supplemental Appropriations Act, 2010 (P.L. 111-212). For more information on the USDA funding, see http://www.or.nrcs.usda.gov/programs/klamath/index.html.

[9] $50 million to aid water conservation efforts in the Klamath was provided in the 2002 farm bill (P.L. 107-171). Additionally, funding under general authorities was provided under both the

2002 farm bill and the 2008 farm bill (P.L. 110-246), although exact amounts are not available.

[10] Although negotiations occurred throughout the early 2000s, many trace the current agreements to a series of administrative hearings in 2006 related to the relicensing of the Klamath Hydroelectric Project under the Federal Power Act (16 U.S.C. §797(e)).

[11] The agreements, as well as other documents related to the Klamath restoration process, are available at http://klamathrestoration.gov/.

[12] This may include, but not be limited to, lease or purchase of water.

[13] 2011 estimates by nonfederal interests noted that approximately $262 million in ongoing "base funding" for Klamath restoration (i.e., funding currently spent under existing authorities) should be available for redirection toward actions in the KBRA, resulting in the lower estimate (approximately $536 million) for "new" federal funding required to implement the KBRA. See Ed Sheets Consulting, "Klamath Basin Restoration Agreement Revised Cost Estimates." June 17, 2011, http://216.119.96.156/Klamath/2011/06/RevisedCost Estimates.pdf. Hereinafter Revised Cost Estimates.

[14] Recommendation of a Dam Removal Entity, or DRE, would be a part of the final Secretarial Determination, which requires congressional authorization.

[15] The dam removal study process included 50 different science, engineering, technical, and economic studies. See http://klamathrestoration.gov/sites/klamathrestoration.gov/files/ 2013%20Updates/Final%20SDOR%20/ 0.Final%20Accessible%20SDOR%2011.8.2012.pdf.

[16] The settlement states that a final Secretarial Determination on dam removal may not be made until federal legislation has been enacted. See, Klamath Hydroelectric Settlement Agreement, p. 20. Available at http://klamathrestoration.gov/ sites/klamathrestoration.gov /files/Klamath-Agreements/Klamath-Hydroelectric-Settlement-Agreement-2-18-10signed.pdf. Accessed May 10, 2011. (Hereinafter Hydroelectric Settlement.)

[17] For example, in addition to its responsibilities under the federal Endangered Species Act and Clean Water Act, as well as other federal laws, the Department of the Interior plays an important role in the potential call for administration of water rights under state law for project irrigators, refuges, and the Klamath Tribe.

[18] DOI has estimated that dam removal itself will create approximately 1,400 jobs in the one-year timeframe for this project, while other actions under the KBRA will create 4,600 jobs over 15 years, with additional gains to farming and fisheries industries See Klamath Regional Economics Fact Sheet, available at http://klamathrestoration.gov/sites/ klamathrestoration. gov/files/Econ.Fact.Sheet.Sept.21.pdf.

[19] Thomas P. Schlosser, "Dewatering Trust Responsibility: the New Klamath River Hydroelectric and Restoration Agreements," Washington Journal of Environmental Law and Policy, vol. 1, no. 1 (July 2011), p. 60. Available at http://digital.law.washington.edu/dspace-law/bitstream/handle/1773.1/1043/1WJELP042.pdf?sequence=1. (Hereinafter Schlosser.)

[20] KBRA, Appendix C-2, p. C.6. Costs for the agreement were estimated in 2007 dollars.

[21] Recent estimates sometimes note that approximately $262 million in ongoing "base funding" for Klamath restoration (i.e., funding currently spent in the Klamath under existing authorities) would be available for redirection toward actions in the KBRA, resulting in a lesser estimate (approximately $536 million) for the required "new" discretionary funding that is needed to implement the KBRA. Revised Cost Estimates, p. 3.

[22] The KHSA provides that the states of California and Oregon are responsible for up to $450 million of the costs for dam removal, but makes no provision for costs beyond this cap. If

estimates conclude that costs are likely to exceed $450 million, then the Secretary must put off a determination until a plan to address these costs is developed.

[23] Department of the Interior, "Final Secretarial Determination Overview Report," October 2012. p. 28. http://klamathrestoration.gov/keep-me-informed/secretarial-determination/role-of-science/secretarial-determinationstudies.

[24] See "Previous Events" section for a breakdown of this funding.

[25] Non-use values were calculated based on regional and national surveys which asked respondents to estimate their willingness to pay for different restoration scenarios

[26] Department of the Interior, Secretarial Overview Report for the Secretary of the Interior, An Assessment of Science and Technical Information, January 23, 2012, p.364 http://klamathrestoration.gov/sites/klamathrestoration.gov/files/ 2013%20Updates/ Final%20SDOR%20/0.Final%20Accessible%20SDOR%2011.8.2012.pdf

[27] National Academy of Science, National Research Council, Endangered and Threatened Fishes in the Klamath River Basin: Causes of Decline and Strategies for Recovery. Washington, DC 2004. p. 81.

[28] The complaint is available at http://prhouser.com/allegation.pdf.

[29] This complaint is available at http://www.peer.org/assets/docs/noaa/ 1_7_13_PEER_Scientific_Misconduct_Complaint.pdf.

[30] The complaint is available at http://media.redding.com/media/static/ Scientific_Integrity_Complaint_20130319.pdf.

[31] Klamath Basin Coordinating Council, "Summary of the Klamath Basin Settlement Agreements," May 2010, available at http://216.119.96.156/Klamath/Summary%20of%20Klamath%20Settlement%20Agreements%204-5- 10.pdf.

[32] For example, see KBRA Sections 2.1, 19.1, 20.3.1, and 22.5.

[33] The 2013 joint biological opinion for the operation of the Klamath Project concluded that the ongoing operation of the project as proposed by Reclamation is not likely to jeopardize the existence of federally listed species. National Marine Fisheries Service, United States Fish and Wildlife Service, Biological Opinions on the Effects of Proposed Klamath Project Operations from May 31, 2013, through March 31, 2023, on Five Federally Listed Threatened and Endangered Species, SWR-2012-9372, Klamath Falls, Oregon, May 2013. http://www.usbr.gov/mp/kbao/docs/ Klamath_Project_Biological_Opinion.pdf.

[34] See KHSA, pp 1-2.

[35] General obligations to support the agreement for non-federal parties is laid out in Part I of the KBRA.

[36] The Upper Klamath Water Users Association, which is also party to the agreements, accounts for approximately half of the off-project acreage. As previously noted, off-project irrigators are not provided similar assurances for water supplies. However, they may stand to benefit from additional supplies and water rights retirement programs included in the KBRA that would not otherwise be available, for example.

[37] As part of the approval process for the ratepayer surcharge for dam removal under the KHSA in Oregon and California, PacifiCorp presented information and received concurrence from state regulatory agencies that its proposed ratepayer increases under the KHSA were fair and reasonable for customers compared to likely costs under relicensing. PacifiCorp estimated that relicensing actions such as construction of fish passage and water quality facilities, as well as reduced flow conditions, would cost in excess of $460 million, and would entail more costs to ratepayers than the approximately $200 million ($172 million in 2010 dollars) pledged under the KHSA.

[38] A previous study by the California Energy Commission and the Bureau of Reclamation found that removal of all dams would be the most cost-effective action for PacifiCorp (i.e., less expensive than modification and ongoing operation of the dams), and some have argued that without the KHSA, the dams would be removed. The final environmental impact statement for FERC relicensing of the project (i.e., before the KHSA was signed) recommended a new license with fish passage and other modifications (see http://www.ferc.gov/industries/hydropower/enviro/eis/ 2007/11-16-07.asp for relicensing documents), which PacifiCorp has argued would be prohibitive (see above note). While some have argued that these costs would eventually force PacifiCorp to surrender its license and fund dam removal itself, PacifiCorp argues that it would pursue relicensing over license surrender and customer-funded dam removal, which have no cap on expenses or liability protections such as those included in the KHSA.

[39] Approximately half of off-project land owners are represented by the Klamath Off-Project Water Users Association, which opposes the agreements. See also U.S. Congress, House Committee on Natural Resources, Subcommittee on Water and Power, Testimony of Thomas Mallams, Klamath Off Project Water Users Association, The Bureau of Reclamation and the American Recovery and Reinvestment Act: A Progress Report and Planning for the Future, Hearing on the American Recovery and Reinvestment Act, 111th Cong., 2nd sess., July 15, 2010.

[40] In its draft EIS, DOI asserts that most of these concerns are unfounded.

[41] This is particularly the case in Siskiyou County, where 79% of voters expressed opposition to removal of the three PacifiCorp dams in California.

[42] In particular, the expert panels associated with Chinook and Coho salmon pointed out uncertainties associated with ongoing water quality issues and have been highlighted by opponents. See for example,

[43] Thomas P. Schlosser, "Dewatering Trust Responsibility: The New Klamath River Hydroelectric and Restoration Agreements," Washington Journal of Environmental Law and Policy, vol. 1, no. 1 (July 2011), p. 42. Available at http://digital.law.washington.edu/dspace-law/bitstream/handle/1773.1/1043/1WJELP042.pdf?sequence=1. (Hereinafter Schlo sser.)

[44] See KBRA, Section 15.4.3.

[45] Ani Kame'enui and Alexander Borack, "Op-ed: Water Quality Suffers as Congress Dithers," Redding Record Searchlight, June 13, 2011.

INDEX

D

E

F

G

H

T

U